# The Green Fiscal Mechanism and Reform for Low Carbon Development

T0227849

This book reviews how far East Asian nations have implemented green fiscal reform, and show how they can advance carbon-energy tax reform to realize low carbon development, with special reference to European policy and experience.

East Asian nations are learning European experiences to adopt them in their political, economic and institutional contexts. However, implementation has been slow in practice, partly due to low acceptability that comes from the same concerns as in Europe, and partly due to weak institutional arrangements for the reform. The slow progress in the revenue side turns our eyes to the expenditure side: how East Asian nations have increased environmental-related expenditures, and how far they have greened sectorial expenditures. This "lifecycle" assessment of fiscal reform, coupled with the assessment of the institutional arrangement, constitutes the features of this book.

The book helps to provide an overall picture of green fiscal reform and carbon-energy tax reform in the East Asian region. The region has a variety of countries, from lowest income to high income nations. Nations have different interests in substance and barriers for reform. This book covers recent development of environmental fiscal reform and carbon-energy taxation in wider nations in the region, including South Korea, Taiwan, Thailand, Vietnam, Indonesia and Japan. In addition, the book's holistic view helps to understand why a specific nation has interest and concern on some aspects of the reforms.

**Akihisa Mori** is an Associate Professor of the Graduate School of Global Environmental Studies, Kyoto University, Japan.

**Paul Ekins** is Professor of Resources and Environmental Policy, and Director of the UCL Institute for Sustainable Resources. He is also a Co-Director of the UK Energy Research Centre, in charge of its Energy Systems theme.

**Soocheol Lee** is a Professor of Economics, Meijo University, Japan.

**Stefan Speck** is an environmental economist with a PhD in economics from Keele University in England.

**Kazuhiro Ueta** is a Professor of Graduate School of Economics, Kyoto University, Japan.

# Routledge studies in ecological economics

# The Green Fiscal Mechanism and Reform for Low Carbon Development

East Asia and Europe

**Edited by Akihisa Mori, Paul Ekins, Soocheol Lee, Stefan Speck and Kazuhiro Ueta**

LONDON AND NEW YORK

First published 2014 by Routledge

2 Park Square, Milton Park, Abingdon, Oxfordshire OX14 4RN
711 Third Avenue, New York, NY 10017

*Routledge is an imprint of the Taylor & Francis Group, an informa business*

First issued in paperback 2018

*British Library Cataloguing in Publication Data*
A catalogue record for this book is available from the British Library

*Library of Congress Cataloging in Publication Data*
The green fiscal mechanism and reform for low carbon development:
East Asia and Europe / edited by Mori, Akihisa, Lee, Soocheol, Ueta,
Kazuhiro, Stefan Speck and Paul Ekins.
    pages cm. – (Routledge studies in ecological economics; 31)
    Includes bibliographical references and index.
    1. Environmental impact charges–East Asia. 2. Environmental impact
    charges–Europe. 3. Environmental policy–Economic aspects–East Asia.
    4. Environmental policy–Economic aspects–Europe. 5. Carbon dioxide
    mitigation–East Asia. 6. Carbon dioxide mitigation–Europe. I. Mori,
    Akihisa, 1970–
    HJ5316.G747 2014
    363.738'7456095–dc23
                                                        2013017244

ISBN: 978-0-415-83956-3 (hbk)
ISBN: 978-1-138-91803-0 (pbk)

Typeset in Times
by Wearset Ltd, Boldon, Tyne and Wear

# Contents

# Figures

# Tables

# Contributors

**Abdurohman** is a researcher at the Indonesian Ministry of Finance, Indonesia.

**Mikael Skou Andersen** is Professor of Environmental Policy Analysis at Aarhus University, Denmark. He has worked on environmental taxation for almost two decades, and has done research more broadly within the field of environmental economics in particular related to externalities. Being originally trained as a political scientist, he has published on policy-making in the European Union too. From 2010–13, he worked with the European Environment Agency, an EU body located in Copenhagen, Denmark.

**Li-chun Chen** is Professor at the Faculty of Economics, Yamaguchi University, Japan.

**Pi Chen** is Assistant Research Fellow at the Center for Energy and Environmental Research, Chung-Hua Institution for Economic Research, Taiwan.

**Paul Ekins** is Professor of Resource and Environmental Policy, Director of the UCL Institute for Sustainable Resources, University College London, and Director of Research at the School of Sustainable Resources and Energy at University College London, England. He has a PhD in economics from the University of London. From 2007–09, he was Director of the UK Green Fiscal Commission. His previous books related to environmental taxation include *Economic Growth and Environmental Sustainability: The Prospects for Green Growth* (Routledge, London, 2000), *Carbon-Energy Taxation: Lessons from Europe* (co-edited with Mikael Skou Andersen, Oxford University Press, Oxford, 2009) and *Environmental Tax Reform: A Policy for Green Growth* (co-edited with Stefan Speck, Oxford University Press, Oxford, 2011).

**Oh Sang Kwon**, PhD in agricultural and resource economics, is Professor of Agricultural Economics and Rural Development at Seoul National University. He is a board member of the East Asian Association of Environmental and Resource Economics. He has written over 90 scholarly journal articles, a book and several book chapters in the areas of welfare analysis of environmental policies, productivity analysis of agricultural, resource and energy sectors, and applied econometric analysis of environmental and food safety issues.

**Soochoal Lee** is Professor of Economics at Meijo University, Japan. He is co-author of "A Model-based Econometric Assessment of Japanese Carbon Tax Reform", *Scientific World Journal*, vol. 2012 (with Hector Pollitt, Kazuhiro Ueta) and "Public Policy Issues on the Disposal of High-level Radioactive Waste in Japan, Critical Issues in Environmental Taxation", Volume XII, *Green Taxation and Environmental Sustainability* (edited by L. Kreiser, A.Y. Sterling, P. Herrera, J.E. Milne, H. Ashiabor, Edward Elgar, 2012 (with Kazuhiro Ueta)) and an editor of *Environmental Charge Systems of East Asian Countries: Features and Challenges* (Showado Press, 2010, in Japanese).

**Zhong Ma** is the Dean and Professor in the School of Environment and Natural Resources, Renmin University of China. He has engaged in national and international environmental policy making including environmental finance and taxation, Montreal Protocol for ODS, $SO_2$ emission trading, and water pricing. He serves as senior advisor to the China Ministry of Environmental Protection (MEP) as well as the provincial governments, and member of the China Council for International Cooperation of Environment and Development (CCICED).

**Yujiao Mao** is a master student at the School of Environment and Natural Resources, Renmin University of China.

**Akihisa Mori** is Associate Professor at the Graduate School of Global Environmental Studies, Kyoto University, Japan. He is an author of *Environmental Aid: Logic, Strategy and Evaluation of Environmental Aid for Sustainable Development* (Yuhikaku, 2009, in Japanese, and Institute for the Environment and Civilization, 2012, in Korean) and editor of *Environmental Governance for Sustainable Development: An East Asian Perspective* (United Nations University Press, 2013) and *Democratization, Decentralization and Environmental Governance in Asia* (Kyoto University Press, 2012).

**Le Thi Kim Oanh** is Vice Rector of Danang University of Technology, University of Danang. She has a PhD in environmental economics at Kyoto University, and is Senior Lecturer in economics and environmental management at Danang University of Technology, Vietnam. She has published a number of papers on pollution control policy and Vietnam's environmental protection charges (in Vietnamese).

**Budy P. Resosudarmo** is Associate Professor in the Arndt-Corden Department of Economics, Australian National University, and the Vice President of the East Asian Association of Environmental and Resource Economics. He published several books on environment and natural resources in Indonesia, including *The Politics and Economics of Indonesia's Natural Resources* (Institute of Southeast Asian Studies, 2005), and *Working with Nature against Poverty* (Institute of Southeast Asian Studies, 2009).

**Chapika Sangkapitux** is a researcher at the Graduate School of Global Environmental Studies, Kyoto University, Japan.

**Ben Shaw** is Head of the Environment Group at the Policy Studies Institute in London. His work at the PSI has included work on environmental tax reform, public attitudes to environmental policy and projects exploring the nature and mechanisms of interaction between research, policy and practice. He coordinates the Sustainable Development Research Network (SDRN) and managed the secretariat for the UK Green Fiscal Commission.

**Daigee Shaw**, a research fellow at the Institute of Economics, Academia Sinica, specializes in economic analysis and policy analysis related to natural resources and pollution control issues. He was President of the Chung-Hua Institution for Economic Research from 2006–11, and currently is serving as President of the East Asian Association of Environmental and Resource Economics, President of the Chinese Regional Science Association-Taiwan. He has published many articles in a number of professional journals, including *American Economic Review*, *Journal of Econometrics*, *Journal of Environmental Economics and Management*, *Environmental Economics and Policy Studies*, *Natural Hazards*, *Energy Policy* and *Risk Analysis*.

**Stefan Speck** is an environmental economist with a PhD in economics and is working as a project manager for environmental economics and policies at the European Environment Agency (EEA) in Copenhagen. Prior to that, he was employed as a senior consultant at Kommunalkredit Public Consulting in Austria and as a senior project scientist at the National Environmental Research Institute/University of Aarhus in Denmark within the EU-funded project "Competitiveness Effects of Environmental Tax Reforms" (COMETR). He also contributed to the research project "Resource Productivity, Environmental Tax Reform and Sustainable Growth in Europe" (PETRE) funded by the Anglo-German Foundation.

**Kazuhiro Ueta** is Professor of the Graduate School of Economics, Kyoto University, Japan, and is an editor of *CDM and Sustainable Development in China: Japanese Perspectives* (Hong Kong University Press, 2012), and co-author of "Measuring Future Dynamics of Genuine Saving with Changes of Population and Technology: Application of an Integrating Assessment Model", *Environment, Development and Sustainability*, 13(4): 703–725, 2011 (with K. Tokimatsu, R. Yamaguchi, M. Sato, R. Yasuoka and M. Nishio) and "Capital Depreciation and Waste Accumulation in Capital-resource Economics", *Applied Economic Letters*, 18(6): 519–522, 2011 (with R. Yamaguchi).

**Jian Wu**, a PhD in Economics, is the Associate Dean and Associate Professor of Environmental Economics at School of Environment and Natural Resources of Renmin University of China. She specializes primarily in economic instruments for pollution control and natural resources management, and institutional/economic analysis on environmental policy, focusing on the field of water, biodiversity and energy. Since 2003, she was engaged in China's environmental taxation and fiscal reform study by working closely with policy makers in this area.

# Preface

*Akihisa Mori, Paul Ekins and Stefan Speck*

Two decades of export-led industrialization and rapid economic growth have turned East Asia into one of the world's growth centers. At the same time, the region has been facing multiple environmental challenges, ranging from natural resource depletion and degradation to environmental pollution and accelerating climate change. To address these challenges, East Asian countries have created various government agencies and administrative organizations and have implemented a number of regulatory measures. Some East Asian countries have attempted to capitalize on scientific knowledge, technologies, institutional arrangements and policies that industrialized countries had employed to enhance effectiveness. Lack of technological, institutional and financial capacity, coupled with weak incentives and insufficient political commitment, however, were some of the major constraints that caused implementation deficits on the ground.

Greening the fiscal system has gained increased attention as a policy instrument that enables a country to complement regulatory measures for curbing environmentally harmful activities, as well as to mobilize domestic resources to meet the financial demands for environmental protection. Growing concerns about climate change and emerging global climate governance in the late 2000s have triggered environmental tax reform (ETR), carbon energy taxation and the emergence of emission trading systems as major components of the global environmental policy reform agenda, while rising global energy prices and the fiscal crisis have forced governments to reduce fuel subsidies. European countries have pioneered the implementation of ETR and carbon energy taxation, whereas East Asian countries have been struggling to instigate such bold reforms. Although the Conference of the Parties of the UN Framework on Climate Change failed to agree on setting concrete greenhouse gas emissions reduction targets by 2020, East Asian countries keep their reform efforts to take a favorable position in the coming multilateral negotiations over post-2020 emissions reduction.

Against this background we believe that it is high time to discuss how East Asian countries have developed environmental fiscal mechanisms and advanced green fiscal reform (GFR) to address these challenges in effective ways.

The initial rationale for this research was to conduct a comparative analysis of the environmental fiscal mechanisms in place, and their effectiveness, within the East Asian region. Up to the first half of the 2000s, a number of research activities on this topic were conducted domestically or on a bilateral basis: the Organisation for Economic Co-operation and Development (OECD), the World Bank, the China Council on International Cooperation for Environment and Development (CCICED) and other bilateral donors hired local researchers as consultants to analyze the state of policy implementation and to shed light on future reform perspectives. Yet none of them has enhanced the regional collaboration among East Asian researchers. This has restricted comparative research on the regional harmonization of environmental policies that the European Union has advanced in the process of integration and enlargement, and has prevented information sharing and comparative analysis among East Asian researchers.

To break this deadlock, Kazuhiro Ueta and Akihisa Mori instigated the formation of a group in 2006 to initiate a pilot research in the project entitled "Economic Development and Environmental Policy in East Asia", granted by the Grant-in-Aid for Scientific Research from the Japan Society for the Promotion of Science (JSPS). They also organized several East Asian expert meetings on this topic to enhance research collaboration among East Asian researchers during 2006–08 as part of research conducted under the Kyoto Sustainability Initiative (KSI) by multiple departments and research institutes of Kyoto University and granted financial support by the Japan Science and Technology Agency (JST). In these pilot research activities and seminars, we acknowledged that the focus of environmental fiscal mechanisms and reform varied considerably among East Asian countries. This led us to expand our research scope to include not only environmental taxation and subsidies conventionally discussed by environmental economists, but also government environmental expenditure and even foreign environmental aid, for the latter finances much of the public environmental expenditure in several countries.

The research underlying this book was made possible with a grant from the Mitsui Bussan Environmental Fund to the project entitled "Environmental Fiscal Mechanism for Sustainable Development in East Asia". The Fund offered financial support not only for promoting this East Asian collaborative research project, but also for establishing and managing the East Asian Association of Environmental and Resource Economics (EAAERE). As such, the research went hand in hand with the development of the Association. In January 2010, we held a preparatory board meeting for the Association, and invited directors and distinguished members of the Association to the kickoff meeting of the research project at the same time. At the First Association Congress in August 2010, we organized special sessions on this issue and invited Paul Ekins as a keynote speaker to give a talk about the European experience on green fiscal reform. At the Second Association Congress in February 2012, some of our research collaborators made presentations on this issue, and we invited Thomas Sterner as a keynote speaker to give a talk about the global experience on fuel tax and subsidy reform, on which Indonesia has focused for several years.

The invitation of Paul Ekins triggered a further expansion of the research scope to include carbon energy tax and European experiences. Several countries in Europe had pioneered green fiscal reform and the implementation of a carbon energy tax, advancing research into overcoming barriers in the process. We found that this experience may have important implications for and relevance to the East Asian countries, even though the development process and contexts may differ considerably. In November 2011, Paul Ekins, Mikael Skou Andersen and Soocheol Lee organized the "Euro-Asia Expert Workshop on Carbon Energy Tax and Green Fiscal Reform in the context of a Sustainable Low Carbon Economy" at University College London (UCL) to share knowledge and present findings of the research. This workshop together with the EAAERE congresses has been a major stimulus for our work.

By evaluating the current status of environmental fiscal mechanisms and exploring current and future feasibility and challenges for carbon energy taxation and green fiscal reform in East Asia, this research, of which this book is an outcome, sought to make a contribution to one of the most important issues facing the region today: how to integrate environmental and climate concerns into national and local fiscal systems in East Asia, referring to experiences and findings in Europe and aligning reform directions with the goal to harmonize environmental fiscal policy regionally and globally.

# Acknowledgments

Our first acknowledgment and grateful thanks go to the Mitsui Bussan Environmental Fund, which funded a four-year research project entitled "Environmental Fiscal Mechanism for Sustainable Development in East Asia", from which all the chapters in Parts I and III in this book are derived. Besides funding the various research activities, the Fund provided financial support to hold congresses of the East Asia Association of Environmental and Resource Economics (EAAERE) at Sapporo, Hanoi and Bandung, and international expert meetings at Kyoto, Beijing and London.

We would like to thank the European Environment Agency and the UCL Energy Institute, University College London, for providing a grant to the "Euro-Asia Expert Workshop on Carbon Energy Tax and Green Fiscal Reform in the context of a Sustainable Low Carbon Economy", from which three chapters in Part II of this book stem.

We would also like to thank Chapika Sangkapitux for editing the chapters in Parts I and III. Her insightful comments on both the contents of the chapters and use of English contributed to improving the quality of this publication. We are grateful for Rumi Kawazoe and Eriko Iida for their managerial services to this research project.

We are grateful to our research collaborators for four years of productive research activities and for the efficiency with which they delivered their chapters for this book.

Finally, our thanks go to the editorial and production staff at Routledge for accepting the publication project and for providing professional assistance toward publishing this book.

Akihisa Mori, Paul Ekins and Stefan Speck
Kyoto, London and Copenhagen
March 2013

# Introduction

*Akihisa Mori, Paul Ekins and Stefan Speck*

## Sustainability, climate change and international responses

### The challenge of environmental sustainability

Sustainability itself simply means that whatever is being considered has the capacity for continuance. Environmental sustainability requires the maintenance of important environmental functions and the natural capital which generates them, where important environmental functions may be considered to be those that are not substitutable, those whose loss is irreversible and is likely to lead to 'immoderate' losses (i.e. costs considerably greater than the costs of maintaining the functions), and those that are crucial for the maintenance of health, for the avoidance of substantial threats (such as climate instability) and for economic sustainability.

Any aspiration for a sustainable economy, or sustainable economic growth, must start from the recognition of the need for the sustainable use of resources and ecosystems, and be rooted in basic laws of physical science, which hold that indefinite physical expansion of the human economy on a finite planet is impossible; and that, with the partial exception of natural, ambient energy, all use of energy creates disorder, and potential disruption, in the natural world. The laws of thermodynamics mean that, at a certain physical scale, further physical growth becomes counter-productive, and there is now substantial evidence that, except from a very short-term perspective, there is little doubt that this scale has now been exceeded. Work by Rockström *et al.* (2009) suggests that this is already the case for biodiversity loss, climate change and the nitrogen cycle, with the phosphorus cycle also fast approaching the limit.

The Rockström analysis is entirely consistent with the more detailed assessments of climate science of IPCC (Intergovernmental Panel on Climate Change 2007) and the Millennium Ecosystem Assessment (MEA 2005), which make clear that without a radical reform of the human–nature relation – in favour of nature – human civilization is at grave threat. Specifically, the evidence strongly suggests that nine billion humans cannot live current Western lifestyles and maintain a habitable planet: the first thing to go will be climate stability, the whole biosphere may then start to unravel. Any aspiration for a sustainable economy must start from the recognition of the need for the sustainable use of resources and ecosystems, rooted in basic laws of physical science.

## The challenge of climate change

The first priority for environmental sustainability, as acknowledged at the 2009 Conference of the Parties to the UN Framework Convention on Climate Change (UNFCCC) is to reduce emissions of greenhouse gases to keep global average warming below 2°C. Figure I.1 shows the emissions reductions required in order to avoid global warming of more than 2°C. The upper shaded areas (broadly those above and immediately around the darker shaded area) show emissions pathways along which there is a 50 per cent probability of achieving this target; the lower shaded areas (including the lower lighter shaded areas) show emissions pathways along which there is a 75 per cent probability of achieving this target. (The various SRES scenarios developed by the IPCC, A1F1 down to B2, which do not allow for any explicit climate change mitigation policies are shown for the purpose of comparison.) For a greater than 50 per cent chance of achieving this, global greenhouse gas emissions will have to fall by around 50 per cent by 2060.

Achieving this kind of reduction in carbon emissions would require carbon and energy intensity (where this is defined as carbon emissions or energy use divided by GDP) to decrease at an unprecedented rate. To illustrate, Jackson

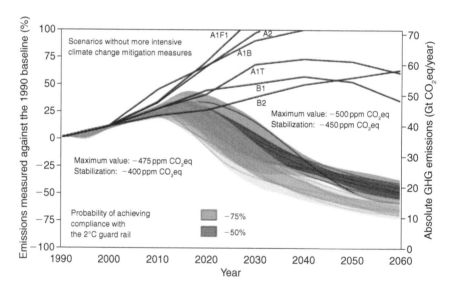

*Figure I.1*    The relationship between greenhouse gas emissions and average global warming (source: WBGU 2007, Figure 8.2.1, p. 165, based on Meinshausen, 2006).

Note
WBGU (German Advisory Council on Global Change) (2007) *Climate Change as a Security Risk*, London: Earthscan.

(2009, pp. 79–80) notes that global carbon intensity over 1990–2008 decreased by 0.7 per cent p.a. on average, but this was overwhelmed by population and economic growth over the same period, so that global emissions grew by 2 per cent p.a. To reach greenhouse gas atmospheric concentrations of 450 ppm (which broadly corresponds to a 50 per cent chance of keeping the global temperature increase to below 2°C (Stern 2007, p. 220)) by 2050, emissions would need to fall by about 5 per cent p.a. With continuing economic growth of 2.1 per cent p.a., carbon intensity would have to fall by around 7 per cent p.a., or about ten times faster than it has since 1990. The policy challenge to achieve these kinds of changes in energy use is unprecedented.

In the analysis of policies to reduce greenhouse gas emissions in the Stern Review (Stern 2007, p. xviii), three kinds of policies were perceived as likely to be required:

- carbon pricing, such as through carbon taxes or emission trading;
- technology policy, to stimulate the rapid development and deployment of low carbon energy sources, high-efficiency end-use appliances and buildings and a large-scale investment programme;
- removal of other barriers to energy efficiency and the promotion of behaviour change to encourage the take-up of new technologies and high-efficiency end-use options, and low-energy (carbon) behaviours (i.e. less driving/flying/meat-eating/lower building temperatures in winter, higher in summer).

The basic economic insight in the Stern Review and elsewhere is that, in a market economy, pricing is the key to resource efficiency, investment and behaviour change. That identifies a key role in climate change and environmental policy for environmental taxation in general and carbon taxation in particular. Many industrialized countries, especially but not only in Europe, have already begun to implement environmental taxation, often in the context of a broader environmental tax reform (ETR). The definition and scope of such a reform is further discussed below. One of the most significant developments in recent years is that this approach to reducing greenhouse gas emissions, and to environmental improvement more generally, is now spreading well beyond the industrialized countries that pioneered its introduction.

### International responses

International society has created a global architecture that stimulates a shift toward a low carbon economy. At the 1997 UN Conference of Parties under the UNFCCC in Kyoto, the parties agreed to legally binding reductions in their emissions of greenhouse gases between 2008 and 2012, followed by the implementation of an International Emissions Trading scheme, Joint Implementation and Clean Development Mechanism (CDM) that provide them with cost-effective emission reduction options.

Climate change mitigation has been recognized as a challenge for and an obligation of high-income countries that used to be the major emitters. In the mid-2000s, however, it was proved that developing countries discharged already half of the annual greenhouse gas emissions. China in particular has rapidly increased its emissions and became the largest annual emitter in 2006, while other emerging economies, such as India, Brazil and South Korea, have shown the same trend. Climate change mitigation therefore has become a major priority on the global agenda.

However, most developing countries have been reluctant to do their part. They have been worried that the imposition of greenhouse gas emissions reduction targets would restrict their future economic growth. They felt it was unfair that high-income countries had enjoyed economic growth for a long period of time with a huge amount of emissions accumulated over their industrialized history, while not allowing developing countries to follow the same growth path. While most developing countries ratified the Kyoto Protocol, they insisted on the principle of common but different responsibility and did not accept any obligations on emission reduction.

Their attitudes changed somewhat when they perceived a clear benefit from the CDM. China, India and Brazil are the largest winners of CDM, for they have gained substantial financial benefits and adopted advanced low carbon technologies from industrialized countries. Developing countries with significant forest areas are also to obtain an opportunity to gain the financial benefits under the Reducing Emissions from Deforestation and Forest Degradation (REDD) mechanism. These global financial mechanisms make it easier for emerging countries to make a firm commitment to greenhouse gas emissions reduction in the post-2012 framework: many of them submitted to the United Nations a list of Nationally Appropriate Mitigation Actions (NAMAs) that show greenhouse gas emissions reduction targets by 2020 (Table I.1). Despite the voluntary nature of these NAMAs, East Asian nations have implemented a variety of policies and measures to attain these targets, which serve as inputs into future international negotiations.

Despite the failure to reach an agreement on the legally binding reduction target in the post-2012 period, the parties decided to make a collective pledge to provide new and additional resources for climate mitigation and adaptation (fast-start finance: FSF) at the 2009 Conference of Parties in Copenhagen. They also agreed to establish the Green Climate Fund (GCF) at the 2010 Conference in Cancun. These new financial arrangements are expected to stimulate developing countries towards pursuing low carbon development.

However, the pledges to FSF fall short of the actual needs, and only 30 per cent of the pledges have been allocated to projects by June 2011 (Stadelmann *et al.* 2012, p. 123). An array of innovative financing options with large fundraising potential has been proposed for financing the GCF and the post-2015 Agenda (e.g. UN Department of Economic and Social Affairs 2012), but it is uncertain whether the GCF will obtain sufficient resources to leverage private funding and ensure wider access to developing countries.

Table 1.1 Mid-term GHG emissions reduction targets

| Nation | Reduction target by 2020 | Term | Base year | Obligation |
|---|---|---|---|---|
| EU | 20% | Absolute volume | 1990 | Commitment |
| | 30% | | | Conditional |
| Japan | 25% | Absolute volume | 1990 | Conditional |
| United States | In the range of 17% | Absolute volume | 2005 | Conditional |
| China | 40–45% (8.5% reduction compared to BAU in 2020) | Per unit of GDP | 2005 | Voluntary |
| India | 20–25% | Per unit of GDP | 2005 | Voluntary |
| Indonesia | 26% when compared with BAU scenario | Absolute volume | 2005 | Voluntary |
| | 41% compared with BAU scenario | | | Voluntary and conditional |
| South Korea | 30% compared with BAU scenario (4% reduction) | Absolute volume | 2005 | Voluntary |
| Taiwan | 0% increase in 2025 | Absolute volume | 2005 | Voluntary |

Source: author's compilation.

This implies that, despite the global climate architecture, domestic resources still constitutes the main source of finance for climate change mitigation action. Environmental fiscal reform is required not only to impose a price on carbon emissions, but to mobilize domestic resources to meet environment and climate needs.

## Environmental tax, green fiscal reform and environmental fiscal mechanism

### *The rationale for environmental taxes*

It is a basic insight of environmental economics, as set out briefly below, that where environmental market failures lead to excessive pollution and environmental destruction, then environmental policy can be justified and, among the instruments of environmental policy, environmental taxation is often more efficient than regulation, and more effective than voluntary agreements and the simple provision of information.

Where the environmental damage comes about because of energy use, there is a special justification for using energy taxes, because energy demand increases with income; energy demand decreases with price; there are market failures for some energy efficiency technologies; improvements in energy efficiency with no increase in energy price can lead to a rebound effect and, therefore, save less energy than anticipated; and humans are extremely ingenious at finding new ways to use energy (heating drives, gardens, making artificial snow, etc.) (Oreszczyn 2004).

It may also be noted in respect of reducing carbon emissions, that only carbon pricing (taxing or trading) will stimulate both the uptake and development of existing low carbon and efficiency technologies, and reduction in the demand for carbon-based fuels.

In summary the conclusions from the literature on environmental taxation are unequivocal:

- without environmental taxation, the (macro-economic) cost of environmental improvement will be higher than it needs to be;
- without significant increases in energy prices, energy consumption will go on rising;
- where the energy is carbon based this will lead to increased carbon emissions and a failure to stabilize the climate.

This book is mainly about the possible role in addressing GHG emissions and other environmental problems of a particular implementation of environmental taxation, called environmental tax reform (ETR) or green fiscal reform (GFR) to bring about the kind of carbon emission reductions that are required by the analysis above.

### *Environmental (or green) tax (or fiscal) reform*

During the last 25 years a special type of a public policy instrument, which is called environmental tax reform (ETR), also known in the literature under the term ecological tax reform, environmental fiscal reform or green fiscal reform (GFR), has been analysed by many economic scholars and institutions but also implemented by countries throughout the world. Although the concept and underlying principle of an ETR is widely endorsed, a generally agreed definition of ETR does not exist.

The European Environment Agency (EEA) has defined ETR as 'a reform of the national tax system where there is a shift of the burden of taxes from conventional taxes such as labour to environmentally damaging activities, such as resource use or pollution' (2005, p. 84). It can therefore be said that an ETR is a tax-shifting policy whereby taxation is shifted from 'goods' such as labour (e.g. income taxes, social security contributions) or capital (e.g. corporation taxes) to 'bads' (pollution, resource depletion), thereby guaranteeing that price signals – as a consequence of introducing environmental taxes – provide an incentive to consumers and producers to change behaviour. The basis for implementing environmental taxes within an ETR is the generally accepted premise that taxes are a tool for reaching political objectives effectively and in a cost-minimizing way.

Slightly broader definitions apply with regard to the model of an environmental fiscal reform (EFR). For example, the Organisation for Economic Co-operation and Development (OECD) describes the concept of an environmental fiscal reform in terms of 'a range of taxation and pricing measures which can raise fiscal revenues while furthering environmental goals' (2005, p. 12). The main difference between these classifications is that an ETR focuses on changing the national tax system, i.e. a tax-shifting policy programme, as compared to an EFR which comprises a wider spectrum of economic measures potentially to be implemented, which beside taxing measures also includes pricing policies in the water and waste area as well as policies aiming to reform subsidies.

Broader still is the concept of 'environmental fiscal mechanism', defined as institutional arrangements for the effective management of state environmental revenue and expenditure. Such arrangements are still lacking in a number of developing countries, including in East Asia, but are necessary for the effective implementation of both ETR and EFR.

Another aspect that deserves some attention is the issue of revenue neutrality. This is sometimes seen as an important characteristic of an ETR concept, in line with the perception of ETR involving a shift, rather than a change, in the overall tax burden. However, the tax-shifting approach and associated revenue neutrality principle is not a component of the underlying EFR concept as articulated by the definition of the World Bank: 'EFR has the potential to free-up economic resources and generate revenues that can help finance poverty reduction measures, for example infrastructure that improves access of the poor to water, sanitation and energy services' (2005, p. ii).

However, while it is not surprising that countries in different situations view ETR differently in terms of both its purpose and its implementation, making it difficult to nail down a common definition in all circumstances, common to all the intentions behind ETR is the objective of the policy programme to generate environmental benefits by reducing environmental pollution and resource use. Achieving this objective crucially depends on the existing and prevailing economic, fiscal, institutional, legal and environmental conditions in the countries concerned, and these conditions influence the selection of economic instruments to be used. The economic instruments mainly discussed in the literature and also implemented as part of an ETR/EFR package are energy and carbon taxes (Speck and Jilkova 2009), which generate by far the largest amount of revenues in developed countries compared to other environmental taxes (Eurostat 2012). What is done with the revenues accruing from environmental taxation depends largely on country-specific economic, political and financial circumstances, and has to be decided by decision makers based on then current political objectives in the countries concerned. This may differ between European and Asian countries, as detailed in this book, because they face widely differing challenges and opportunities.

### *Theoretical basis of GFR*

Pigou (1920) provided the theoretical case for the application of environmental taxation as environmental taxes can be used to correct for market-based negative externalities. The economic literature characterizes a negative externality, for example GHG emissions or air pollution, such as $SO_2$ emissions, when the price to the consumer of a good or service is less than the cost of that good or service to society as a whole. A tax can internalize the negative externalities as shown in Figure I.2.

The optimal tax rate leads to an efficient allocation of pollution and this is reached when the marginal benefit of pollution intersects the marginal social cost. Without this intervention the market would be at equilibrium where marginal benefit intersects the marginal private cost, i.e. an inefficient allocation of pollution (Speck *et al.* 2011). The identification of the optimal level of

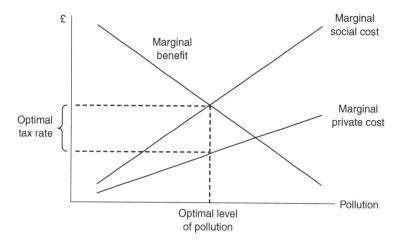

*Figure I.2* The efficient allocation of pollution (source: Speck *et al.* 2011).

pollution is not a straightforward task at all, implying that the Pigouvian taxation approach encounters real difficulties in the real world. However, Baumol and Oates (1971) proposed an alternative approach to implementing environmental taxes which does not require the determination of the optimal tax, which is known as the standards and pricing approach in the economic literature. The starting point of this approach is to choose environmental standards and then to use environmental taxes on an iterative basis to bring levels of environmental pollution down to these predetermined standards. Current policies generally follow this approach by introducing new environmental taxes or by increasing tax rates, thereby achieving the policy objective of creating environmental benefits by reducing environmental pollution and resource use.

As discussed above the concept of an ETR is not solely directed to the application of environmental taxes but also involves recycling the revenues back to the economy. The widely accepted form of recycling is to reduce other distortionary taxation, such as labour taxes, and thereby to increase employment and welfare, i.e. the possibility of achieving a 'double dividend' (Oates 1995).

The double dividend hypothesis is heavily discussed and contested in the economic literature. For example, Goulder (1995) was one of the first scholars analysing the concept in some detail and splitting it into a weak and strong version. The weak version does not lead to a great controversy as it claims that the costs of environmental taxation will be reduced if the revenue is recycled into reducing existing tax distortions. In contrast, the strong version of the double dividend concept is heavily debated as it states that the process of revenue recycling will actually lead to an improvement in welfare as employment and/or economic output will increase. The debate surrounding the existence of the strong double dividend hypothesis is still ongoing and not further discussed in this book. However, the theoretical discussion clearly reveals that the prevailing fiscal system as well as the adopted revenue recycling mechanism as part of an ETR is crucial for answering the question whether the double dividend exists (Schöb 1996; Bovenberg and van der Ploeg 1998; Heady *et al.* 2000).

Distributional implications and the potential loss of competitiveness of industries are the regularly mentioned obstacles for the more widespread application of ETRs. Both aspects are assessed in the economic literature in detail and cannot be ignored as environmental taxes, in particular energy and carbon taxes, lead to higher energy prices and if introduced unilaterally can impair the competitiveness of domestic industry. However, these concerns are accommodated in the political reality of designing ETRs. The distributional implications can be and are addressed by adopting revenue recycling measures targeting these equity effects (EEA 2011a).

The competitiveness discussion attracts even more attention in the economic literature (Ekins and Speck 1999, 2008; Andersen and Ekins 2009; EEA 2011b). The main argument put forward against an ETR is the increase in the costs of energy inputs as a consequence of the unilateral introduction of higher energy and carbon prices. However, this argument is disputed by the Porter Hypothesis (Porter and van der Linde 1995) arguing that any form of environmental regulation including environmental taxation can lead to an increase in efficiency and innovation efforts so that the competitiveness of the affected industries is improved over time as these efficiency gains offset the costs of complying with the regulation. Furthermore, industrial sectors and in particular energy-intensive sectors are regularly facing lower energy and carbon tax rates because of these concerns (Speck and Jilkova 2009). It can be stated that these constraints are real and must be taken seriously but they should not hinder the application of green fiscal reforms as these negative and damaging consequences of environmental taxes, in particular energy and carbon taxes, can be overcome as the political reality clearly demonstrates.

### The wider scope of GFR

ETR can stimulate a shift of tax structure from labour and capital towards environmental 'bads', thus contributing to greening the revenue side of state budgets. Green fiscal reform, through removal of subsidies on energy, water and pesticides, has implications for greening the expenditure side. The environmental benefits of greening the state budget can be offset, however, if governments continue to spend considerable sums on environmentally harmful activities. In countries where bureaucratic segregation and sectionalism prevail and economic growth is posited as a high priority, sectoral ministries, which are often given higher status in the bureaucratic hierarchy, demand budgets to implement policies, programmes and projects with little consideration for the environment.

To make GFR effective for shifting gear towards a low carbon economy, it becomes necessary to conduct *ex ante* appraisals of government expenditure from a sustainability point of view. Given that many countries employ mid-term sectoral and fiscal plans, it is desirable to expand the scope of sustainability appraisal also to these plans to influence priority setting in fiscal expenditure. *Ex post* monitoring and evaluation is also essential to provide feedback for drafting the next mid-term plans. These measures enable governments to establish the green budgeting 'life cycle' that includes both the expenditure and revenue sides (Figure I.3).

Focusing on the expenditure side is of particular importance in the context of emerging economies where the state revenue is increasing, thus the governments can shift the structure of expenditure much more easily than in developed countries. This is the major reason why we do not limit the scope of this book to ETR and GFR, but have opted for a broader scope that includes the environmental fiscal mechanism.

## Experiences of ETR/EFR/GFR and carbon energy taxation

### Experiences in Europe

Significant tax and fiscal reforms have been implemented in European countries mainly in the Scandinavian countries during the early 1990s and their main objectives have been to reduce labour costs, to broaden the overall tax base leading to an increase in general consumption taxes, such as VAT and

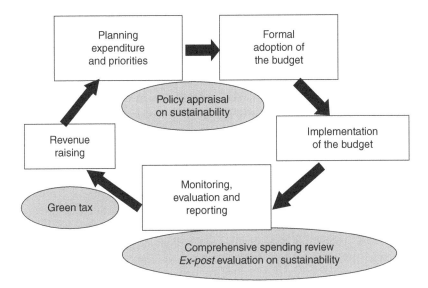

*Figure I.3* The green budgeting 'life cycle' (source: Author compilation on Wilkinson *et al.* 2008).

environmental taxes. The Scandinavian governments adopted the concept of an ETR as part of the much wider fiscal reform. For example, the main policy objective of the Swedish fiscal reform in 1991 was the reduction of personal income taxes. The shortfall in the state budget as a consequence of cutting the income tax rates was partly compensated by levying the value added tax (VAT) on energy purchases and by introducing the $SO_2$ and $CO_2$ tax.

During the 1990s the governments of the Netherlands, the UK and Germany followed. The policy approach of the latter countries was different as they adopted a 'pure' ETR by implementing new environmental taxes, in the majority of the case taxes levied on the consumption of energy and $CO_2$ emissions, and to increasing already existing ones. The other component of the tax-shifting programmes was mainly done by reducing income taxes as well as non-wage labour costs, such as social security contributions, and by raising the tax-free allowances with regard to incomes. All these measures were done adhering to the revenue-neutral principle meaning that the overall tax burden remained constant. However, a small part of the additional revenues were also used for fiscal reduction policies by the German government (Speck and Jilkova 2009).

Other European countries implemented ETRs in their national reform programmes during the last ten years. For example, Estonia enacted an ETR package to be implemented in two phases during 2006 and 2013 by shifting the tax burden from income to taxes and charges levied on energy, natural resources and environmental pollution (Lelumees 2007).

In 2008, Switzerland introduced a $CO_2$ tax levied on fuels – exempting transport fuels – with revenues of the $CO_2$ tax recycled back to companies and households. The novelty of the Swiss $CO_2$ tax package is that the recycling mechanism for household differs from the experiences gained in other European countries as households receive a lump sum with the clearly announced policy objective of reducing the health insurance premium payable by the individual citizen by this fixed amount. The effectiveness of this recycling mechanism can be questioned, in particular when considering the theoretical arguments of the double dividend hypothesis. However, this form of recycling mechanism has an advantage which cannot be underrated as it undoubtedly demonstrates that the $CO_2$ tax revenues are completely recycled back to the citizens and it can therefore increase the political credibility of the ETR (Speck *et al*. 2011).

However, ETRs are nowadays a policy tool in countries outside Europe too. For example, the province British Columbia (BC) in Canada implemented a $CO_2$ tax in July 2008. The introduction of this tax was done in a revenue-neutral way as all revenues are recycled back to the economy via the reduction of personal and corporate income taxes (Speck 2013). In addition, the BC government designed the ETR programme in a way so that specific distributional implications are mitigated by enacting fiscal measures to protect vulnerable households (Harrison 2012). The Australian government also included clearly defined policy measures in its carbon tax legislation, i.e. the Clean Energy Bill in February 2011, which are aimed to support poor families by increasing the tax-free threshold for income taxes and to boost pension and family tax benefits. The Australian ETR package does not ensure revenue neutrality as parts of the revenues are returned back to industries by providing incentives to invest in cleaner energy programmes (Speck 2013). These two country examples demonstrate that ETR packages are shaped such that they address the concerns of citizen and industries, even if thereby they miss out some of the potential gains in terms of increasing effectiveness and improving efficiency in environmental policies, in order to gain increased public acceptance of this policy tool.

An ETR is a policy approach not only for developed but also for developing countries. For example, the National Treasury of South Africa published two reports (2003 and 2006) which are by and large following the framework of the ETR/EFR concept (Speck 2010). The overall ETR framework was also transposed into political reality, as the Budget 2009/2010 executed reduction in the personal income tax rates and stated that the revenue losses are offset by an increase in environmental taxation.

### Experiences and challenges in East Asia

In East Asia, increasing state revenue has enabled the government to increase environmental expenditure in the general budget. Several East Asian countries created specialized funds that imposed energy charges and/or were earmarked for environmental investment. Some of them have increased the environmental budget for the sectoral ministries to advance environmental and/or climate policy integration at the 'source' sectors. Local governments have also increased their expenditure to develop environmental infrastructures and to provide environmental services, as well as to develop capacity to manage the environmental problems in the jurisdictions.

However, governments do not automatically secure increases in environmental expenditure. They may increase the spending when environmental movements become politically sensitive, while easily giving in to political, economical and institutional barriers whenever they perceive less pressure from environmental groups and/or stop making firm commitment. Abuse, fraud and inefficiency in the spending are also factors that reduce political and social support for environmental expenditures. The traditional engineering approach of 'predict and provide', bureaucratic segregation and earmarking may lead to over-investment relative to priority needs in the same sector or across the sectors, resulting in inefficient use and high operational and maintenance costs in the long run.[1]

EFR has gradually been perceived as a key policy instrument for greening growth and attaining the Millennium Development Goals (MDGs). Since the agreement on the slogan 'move beyond the sustainable development rhetoric and pursue a path of green growth' (UN ESCAP 2008, p. 3) at the Fifth Ministerial Conference on Environment and Development in 2005, South Korea and the United Nations Economic and Social Commission for Asia and the Pacific (UN ESCAP) have advocated green growth and lobbied governments to adopt this policy instrument. OECD insists that *EFR can bring not only fiscal and environmental benefits but also poverty reduction benefits directly by helping address environmental problems and indirectly by generating or freeing up resources for anti-poverty programmes* (2005, p. 12).

With these external driving forces, several East Asian countries have gradually taken a step towards EFR and carbon-energy taxation. As demonstrated by the chapters in this book, Japan, South Korea, China and Taiwan have already imposed several types of charges, levies and taxes, while Vietnam is carrying out reforms and Indonesia is working towards the removal of its fuel subsidy.

Nonetheless, the scope and impacts of GFR in the region have been quite limited to date. East Asian countries' concerns regarding competitiveness and equity are no less strong than in Europe. The East Asian region is well known for its system of the 'development state' or 'authoritarian development', where the state places highest priority on economic growth as the national interest and mobilizes financial and human resources to enhance national power. Some East Asian countries have succeeded in catching up with industrialized countries by attracting foreign direct investment and promoting export-led industrialization to enhance international competitiveness in the global market. The recent rapid increase in China's competitiveness has threatened not only its traditional competitors among ASEAN member states, but also South Korea and Japan. This leads to the strong opposition of, say, the Japanese industrial association against carbon-energy taxation and a carbon emission trading system by threatening industrial relocation to foreign countries at the cost of domestic employment, as long as China has no legal obligation to reduce carbon emissions. The South Korean government has kept the electricity price for industrial use lower amid the global energy price hike and domestic inflation, at a cost of €1.5 billion to the Korean Electricity Power Company (KEPCO).

Equity or distributional impacts can be more severe in East Asian countries that have lower average incomes and immense poverty problems. OECD (2005) points out that pricing the environment encourages the rich, who are already connected to water and electricity services, to make more rational consumption choices, but it makes no impact on those who are not connected. The impact of pricing the environment may go beyond increases in consumers' payments. It can deprive the poor and vulnerable groups of *access* to basic infrastructure services such as water, sanitation and energy, which may be vital for their livelihoods. This is especially true in a society where environmental goods, services and assets are perceived as non-economic and/or the traditional entitlement of local communities. OECD (2005) suggests taking complementary measures to those affected. In reality, however, the government does not even take a targeted measures to mitigate adverse impacts and simply tends to add the revenue to the general budget. Even if it attempts to take measures, insufficient administrative capacity may prevent the government from doing so effectively.

Aside from competitiveness and equity, East Asian countries face additional challenges that need to be addressed in implementing GFR. The first one is insufficient capacity of the government. Weak tax administration, coupled with fraud and corruption has led to widespread tax evasion and arrears. East Asian governments had little knowledge of ways to obtain social acceptance for and to administer environmental pricing, with the exception of South Korea and China where governments have historically collected a variety of fees and charges. Even the South Korean government initially chose standards and regulations as its sole policy measures, leaving environmental degradation unsolved until people staged massive protests to call for urgent measures that provided positive and visible changes to the environment.

The second challenge is revenue neutrality. As in other developing countries, many East Asian governments face challenges in raising revenue. A large share of economic activities is conducted by small enterprises and the informal sector, which makes it difficult to collect labour, corporate and sales taxes. Many of them thus have relied on taxing international trade and land, which can be collected more easily. Yet such revenues may need to be phased out amid trade liberalization. Increasing number of countries have obtained corporate income tax, but either offering a number of tax privileges to attract foreign direct investment and promote exports in their industrial development strategy, or removing such privileges to allow a reduction of the corporate tax rate. All of these have prevented them from obtaining tax revenue commensurate with economic growth. In addition, the social security system is often under development, and they have not yet imposed higher charges to fund social pensions and public health insurance. East Asian countries have therefore not much opportunity to reduce labour taxes, corporate income tax or social security levies as a counterpart to implementing environmental taxes and/ or reducing environmentally harmful subsidies. Rather, they place their priority on the fiscal benefits from EFR when they become serious about environmental protection. This adversely affects the political feasibility of GFR.

The third challenge is the introduction of the polluter pays principle. In the transition economies of China and Vietnam, state enterprises pay their surplus to the government and the government finances the investments of state enterprises, including environmental investments. Under this relation, the pollution levy and revenue recycling are merely financial transfers within the government sector, and state enterprises do not experience the incentive effects and sense of shared responsibility. GFR may not bring any fiscal and environmental benefits under this circumstance.

These additional barriers lead us to the assumption that benefits that are expected to be generated by GFR depend at least partially on the development of environmental fiscal mechanism, defined as already noted as institutional arrangements for the effective management of state environmental revenue and expenditure.

Given this assumption, this book aims to address three key questions.

1   To what extent have East Asian countries developed green fiscal mechanism, which constitutes an institutional basis for GFR and carbon-energy taxation?
2   What are the achievements and remaining challenges of GFR and carbon-energy taxation in Europe, and what are the lessons for East Asia?
3   To what extent have East Asian countries advanced GFR and carbon-energy taxation, and how have they overcome barriers?

## Outline of the book

To answer these questions, this book is divided into three parts that consist of 14 chapters, in addition to this Introduction, and a Conclusion.

Part I of this book comprises five chapters that analyse the historical evolution and challenges of green fiscal mechanism in East Asia. In Chapter 1, Akihisa Mori provides an overview that covers the evolution of green fiscal mechanism and its reform in the East Asian countries. It also examines how the global energy price hike and emergence of global climate governance in the late 2000s have affected their decision on green fiscal reform. Chapters 2 and 3 pay special attention to China. In Chapter 2, Zhong Ma and Jian Wu focus on government environmental expenditure in China, which has been one of the major environmental policy instruments. They review the development of the mechanism in the fiscal policy to evaluate the government's priorities in relation to environmental protection. Jian Wu, Yujiao Mao and Zhong Ma in Chapter 3 focus on environmental-related taxes in China. OECD (2010) defines environmentally related taxes as any compulsory, unrequited payment to general government levied on tax bases deemed to be of particular environmental relevance, and includes taxes on energy products, motor vehicles, waste, measured or estimated emissions, and natural resources in this category. However, China, as well as South Korea, imposes a variety of environmental charges, levies and fees that do not fall into this category. Assuming that the OECD's definition leads to underestimating the amount of environmentally related taxes in both countries, Wu, Mao and Ma make a comparative analysis of environmentally related taxes and their revenues between OECD countries and China.

Li-Chun Chen in Chapter 4 describes struggles over a pollution control fee in Taiwan. As in other Asian countries, Taiwan employed the standards and regulation approach to pollution, and increased government environmental expenditure in the 1990s. Faced with the ineffectiveness of this approach, it attempted to implement pollution control fees – environmentally related taxes in a broader definition than the OECD's – in enhancing air and water pollution control and promoting recycling. Chen takes the air pollution control fee and examines the impacts on fiscal revenue and emission reduction.

In Chapter 5, Le Thi Kim Oanh and Akihisa Mori shed light on the recent development of environmental fiscal mechanism in the central and local governments in Vietnam. They find that the Communist Party's serious commitment has prompted both central and local governments to increase environmental expenditures. They find that local governments have also created environmental funds to spend much more than the central ones, while the amount varies among them.

Part II reviews the experiences gained with environmental tax reforms in Europe during the last two decades. These experiences are discussed in four chapters. Paul Ekins reviews in Chapter 6 the most recent research undertaken in Europe with regard to ETR. Based on these findings he provides some broad guidance as to how East Asia might proceed with environmental tax reforms, while clearly stating that the European experiences cannot be directly transferred to East Asian countries.

In Chapter 7, Stefan Speck puts environmental taxation in the context of recent economic and financial developments in Europe, noting that politicians are today looking for policy responses that can contribute to the fiscal consolidation process as well as addressing the considerable environmental and climate challenges facing society. Interest in environmental policies, in particular environmental taxes and trading schemes, is arguably increasing because of their potential to perform this dual role, simultaneously reducing budget deficits and promoting the integration of economic and environmental considerations via a sustainable or green fiscal consolidation process.

Mikael Skou Andersen starts Chapter 8 with a review of the public perceptions of environmental taxation followed by an assessment of recent trends in environmentally related taxation in the European Union (EU) member states. His chapter then goes on to explore the underlying factors explaining the paradox that a tax instrument which was in fact relatively popular with the electorate came to decline in revenue significance.

Chapter 9 is written by Paul Ekins and Ben Shaw and they present the findings from a review of past and present green tax (or fiscal) commissions in

Europe, and from their experience distil suggestions for the effective establishment and operation of such commissions in the future. This chapter includes a discussion of lessons learned from previous commissions as well as a detailed debate on recommendations for future ones.

Part III consists of five chapters that analyse struggles in implementing GFR and carbon-energy taxation in East Asian countries, with special focus on Japan, South Korea, Taiwan, Thailand and Indonesia. Soocheol Lee and Kazuhiro Ueta in Chapter 10 evaluate Japan's energy taxes and their earmarking for infrastructure development to develop a proposal for reform.

The next two chapters focus on the tradeoff between economic efficiency, equity, and GDP growth and job creation in carbon-energy taxation. In Chapter 11, Oh Sang Kwon describes energy pricing and environmental charges, to show that the current multipurpose energy tax induces tax resistance in South Korean society. He then conducts a simulation analysis to compare distortions between current energy tax and taxes that are consistent with the carbon contents of energy products. In Chapter 12, Daigee Shaw and Pi Chen make a simulation analysis to highlight the above tradeoff as a barrier to carbon-energy taxation in Taiwan.

The last two chapters shift the focus to the two Southeast Asian countries of Thailand and Indonesia. Chapika Sangkapitux and Akihisa Mori in Chapter 13 review the development of environmental fiscal mechanism in Thailand and examine factors that have hindered both environmental fiscal reform and the introduction of environmental charges and fees, despite over a decade of political and societal debate. In Chapter 14, Resosudarmo and Abdurohman review the recent development of green fiscal reform in the fossil fuel-rich country Indonesia and lay out the challenges in this country of fiscal policies that seek to promote low carbon development.

We conclude the book by synthesizing arguments made in the previous chapters and teasing out future challenges of green fiscal mechanisms and reform for low carbon development in East Asia.

## Note

1 In Japan, for example, the Ministry of Construction initiated the development of a river-basin sewerage system and attempted to expand it to suburban areas so as to enjoy benefits from economies of scale. This has claimed huge investments on the pipe system while preventing more cost-effective treatment plants, which were initiated by other ministries, from diffusing in rural areas. This approach is too institutionally embedded to be changed towards integrated water resource management.

## References

Andersen, M.S. and Ekins, P. (2009) *Carbon-Energy Taxation: Lessons from Europe*, Oxford University Press, Oxford.

Baumol, W. and Oates, W. (1971) 'The use of standards and prices for the protection of the environment', *Swedish Journal of Economics*, 73(1): 42–54.

Bovenberg, L. and van der Ploeg, F. (1998) 'Tax reform, structural unemployment and the environment', *Scandinavian Journal of Economics*, 100(3): 593–610.

Ekins, P. and Speck, S. (1999) 'Competitiveness and exemptions from environmental taxes in Europe,' *Environmental and Resource Economics*, 13(4): 369–396.

Ekins, P. and Speck, S. (2008) 'Environmental tax reform in Europe: energy tax rates and competitiveness', in: L. Kaiser, J. Milne and N. Chalifour (eds), *Critical Issues in Environmental Taxation: Volume V*, Oxford University Press: Oxford, 77–105.

Ekins, P. and Speck, S. (2011) *Environmental Tax Reform (ETR): A Policy for Green Growth*, Oxford University Press: Oxford.

European Environment Agency (EEA) (2005) 'Market based instruments in environmental policy in Europe', EEA Technical Report No. 8/2005, Copenhagen.

European Environment Agency (EEA) (2011) 'Environmental tax reform in Europe: implications income distribution', EEA Technical Report No. 16/2011, Copenhagen.

European Environment Agency (EEA) (2011) 'Environmental tax reform in Europe: opportunities for eco-innovation implications income distribution', EEA Technical Report No. 17/2011, Copenhagen.

Eurostat (2012) 'Taxation trends in the European Union Data for the EU Member States, Iceland and Norway, 2012 edition', Report prepared by Eurostat and European Commission, Taxation and Customs Union, Luxembourg, Publication Office of the European Union.

Goulder, L. (1995) 'Environmental taxation and the "double dividend": a reader's guide', *International Tax and Public Finance*, 2(2): 157–183.

Harrison, K. (2012) *The Political Economy of British Columbia's Carbon Tax*, Centre for Tax Policy and Administration Environment Directorate, OECD, COM/ENV/EPOC/CTPA/CFA(2012)42, Paris.

Heady, C.J., Markandya, A., Blyth, W., Collingwood, J. and Taylor, P.G. (2000) 'Study on the relationship between environmental/energy taxation and employment creation'. Online, available at: http://ec.europa.eu/environment/enveco/taxation/pdf/entaxemp.pdf (accessed 9 February 2013).

Intergovernmental Panel on Climate Change (IPCC) (2007) *Climate Change 2007: The Physical Science Basis*, Contribution of Working Group I to the Fourth Assessment Report of the Intergovernmental Panel on Climate Change, Cambridge University Press, Cambridge.

Jackson, T. (2009) *Prosperity without Growth: Economics for a Finite Planet*, Earthscan, London.

Lelumees, L. (2007) 'Carrying out the ecological tax reform in Estonia', Online, available at: www.fin.ee/index.php?id=621 (accessed 19 September 2007).

Meinshausen, M. (2006) 'What does a 2°C target mean for greenhouse gas concentrations? A brief analysis based on multi-gas emission pathways and several climate sensitivity uncertainty estimates', in: H.-J. Schellnhuber, W. Cramer, N. Nakicenovic, T.M.L. Wigley and G. Yohe (eds) *Avoiding Dangerous Climate Change*, Cambridge University Press, Cambridge/New York, 265–279.

Millennium Ecosystem Assessment (MEA) (2005) *Ecosystems and Human Well-being: Synthesis*, Island Press, Washington, DC.

National Treasury (2003) 'Market-based instrument to support environmental fiscal reform in South Africa: a discussion document', Pretoria, South Africa.

National Treasury (2006) 'A framework for considering market-based instrument to support environmental fiscal reform in South Africa', Draft Policy Paper, Tax Policy Chief Directorate, Pretoria, South Africa.

Oates, W.E. (1995) 'Green taxes: can we protect the environment and improve the tax system at the same time?' *Southern Economic Journal*, 61(4): 914–922.

Organisation for Economic Co-operation and Development (OECD) (2005) *Environmental Fiscal Reform for Poverty Reduction*, DAC Guidelines and Reference Series, DAC Guidelines and Reference Series, OECD: Paris.

Organisation for Economic Co-operation and Development (OECD) (2010) *Taxes, Innovation and the Environment*, OECD: Paris.

Oreszczyn, T. (2004) 'Our innate ability to think of new ways to use energy', *Energy Policy*, 15(6): 1011–1014.

Pigou, A.C. (1920) *The Economics of Welfare*, Macmillan, London.

Porter, M.E. and van der Linde, C. (1995) 'Towards a new conception of the environment-competitiveness relationship', *Journal of Economic Perspectives*, 9(4): 119–132.

Rockström, J., Steffen, W., Noone, K., Persson, Å., Chapin III, F.S., Lambin, E.F., Lenton, T.M., Scheffer, M., Folke, C., Schellnhuber, H.J., Nykvist, B., de Wit, C.A., Hughes, S., van der Leeuw, S., Rodhe, H., Sörlin, S., Snyder, P.K., Costanza, R., Svedin, U., Falkenmark, M., Karlberg, L., Corell, R.W., Fabry, V.J., Hansen, J., Walker, B., Liverman, D., Richardson, K., Crutzen, P. and Foley, J.A. (2009) 'A safe operating space for humanity', *Nature*, 461 (24 September): 472–475.

Schöb, R. (1996) 'Evaluating tax reform in the presence of externalities', *Oxford Economic Papers*, 48(4): 537–555.

Stadelmann, M., Brown, J. and Hörnlein, L. (2012) 'Fast-start finance: scattered governance, information and programmes', in: A. Michaelowa (ed.) *Carbon Markets or Climate Finance? Low Carbon and Adaptation Investment Choices for the Developing World*, Routledge: London, 117–145.

Speck, S. (2010) 'Options for promoting environmental fiscal reform in EC development cooperation: country case study South Africa', final report of the EC funded project 'Options for Promoting Environmental Fiscal Reform in EC Development Cooperation'. Online, available at: www.unpei.org/PDF/budgetingfinancing/southafrica-case-study-fiscalreforms.pdf (accessed 10 February 2013).

Speck, S. (2013) 'Carbon taxation: two decades of experience and future prospects', *Carbon Management* (forthcoming).

Speck, S. and Jilkova, J. (2009) 'Design of environmental tax reforms in Europe', in: M.S. Andersen and P. Ekins (eds) *Carbon-Energy Taxation: Lessons from Europe*, Oxford University Press, Oxford, 24–52.

Speck, S., Summerton, P., Lee, D. and Wiebe, K.S. (2011) 'Environmental taxes and ETRs in Europe: the current situation and a review of the modelling literature', in: P. Ekins and S. Speck (eds) *Environmental Tax Reform: A Policy for Green Growth*, Oxford University Press, Oxford, 99–130.

Stern, N. (2007) *The Economics of Climate Change: The Stern Review*, Cambridge University Press, Cambridge.

UN Department of Economic and Social Affairs (2012) *World Economic and Social Survey 2012: In Search of New Development Finance*, United Nations, New York.

UN ESCAP (2008) *Greening Growth in Asia and the Pacific*, United Nations, Bangkok.

Wilkinson, D., Benson, D. and Jordan, A. (2008) 'Green budgeting', in: A.J. Jordan and A. Lenschow (eds) *Innovation in Environmental Policy? Integrating the Environment for Sustainability*, Edward Elgar: Cheltenham, 70–92.

World Bank (2005) *Environmental Fiscal Reform What Should be Done and How to Achieve it*, World Bank: Washington, DC.

# Part I

# Environmental fiscal mechanism and reform in East Asia

# 1 Environmental fiscal mechanism and reform in East Asia

## An overview

*Akihisa Mori*

## Introduction

East Asia is diverse in people, religions, climate, incomes, development stage and pathways, resource endowment and political systems. Diversity becomes even larger when both Northeast and Southeast Asia are included.

On the other hand, the entire region has employed the same development strategy: attracting foreign direct investment and promoting export-led industrialization to enhance international competitiveness in the global market. This development mode, which has intensified amid enhanced international division of labor, has often caused similar levels of social and environmental degradation, which raised massive environmental movements and protests, and even brought several high-profile cases to the court.

Emerging environmental problems and challenges require governments in East Asia to implement more stringent environmental laws and policies. Governments have also developed environmental fiscal mechanisms to develop capacity to implement these laws and policies, as well as to finance the development of wastewater and solid waste management system. However, progress varies considerably among countries. Some countries have allocated plenty of fiscal resources not only from their general budget, but also earmarked funds and environmental taxes and charges, while others have done little to secure such fiscal means.

Recent global energy hikes and emerging global climate governance may provide a good reason to implement energy and carbon taxes and to reduce fuel subsidies. East Asian policy makers can refer to European experiences to obtain political and social acceptance to implement reform. Market integration momentum for the ASEAN region may offer a good chance to harmonize environmental policies, including environmental tax and subsidy regionally, because it enables to ensure equal and easier access to the regional market.

There are a number of scenarios and empirical analyses on environmental tax reform in each specific country in East Asia as more and more countries have attempted to implement it. While Japanese experiences were revisited (see, Inui *et al*. 2005 and Mori *et al*. 2005) and simulation analyses of carbon tax policy proposals have been conducted, recent attention has shifted to China, which has suffered from serious environmental pollution and quested for innovative policy instruments. OECD (1999) is one of the earliest analyses, followed by Wang *et al.* (2003), CCICED (2005, 2009), Wang (2011) and Wang *et al.* (2011). A number of international comparative analyses have been conducted, e.g., on the evolution of environmental fiscal mechanism (Ueta *et al.* 2009), environmental charges (Lee 2010) and carbon tax policy (Liu *et al.* 2011) among East Asian countries. However, these are either preliminary or analysis of specific charges, and focus on Northeast Asia, partly due to the significant differences that exist with regard to the definition and range of state environmental expenditure and tax among countries.

Recent studies go so far as to analyze economic and environmental impacts of region-wide environmental tax (Park *et al.* 2012) and carbon emission trading scheme (Ban and Fujikawa 2013).

Even if such reforms provide economic and environmental benefits region-ally, however, how realistic is it to initiate environmental fiscal reform region-ally, let alone to impose minimum common environmental and/or carbon tax rate in East Asia with its high diversity in per capita income and development stage? To answer this question, we need to examine impacts of major driving forces and their limitation.

This chapter takes a closer look at the possible outcomes of environmental fiscal reform and conducts a comparative analysis of the historical development to clarify the specific features. It will target both Northeast and Southeast Asia, including Japan, South Korea, Taiwan, China, Thailand, Indonesia, and Vietnam. Based on the findings, it will discuss the future of environmental fiscal reform in East Asia.

## Development of environmental fiscal mechanism

### *Features*

There are a number of similarities in the development of environmental policies in East Asia. Governments allocate funds to develop environmental infrastructure such as wastewater and solid waste treatment system, to recover polluted areas, to pay compensation to victims, and to offer economic incentives for emission reduction. They have designed environmental fiscal mechanisms for this purpose.

While the definition and scope differs among countries, we can find several commonalities from the trend of state environmental expenditure. First, governments boost their environmental expenditures when they initiate industrial pollution control. Japan rapidly increased the expenditure in the early 1970s, South Korea, Taiwan, Indonesia and Thailand in the 1990s, China after the latter half of the 1990s, and Vietnam in the mid-2000s (Figure 1.1).

Second, Northeast Asian countries have spent a relatively higher amount for the environment than Southeast Asian ones. The ratio of public environmental expenditure/investment to GDP went beyond 1 percent in Japan and China,

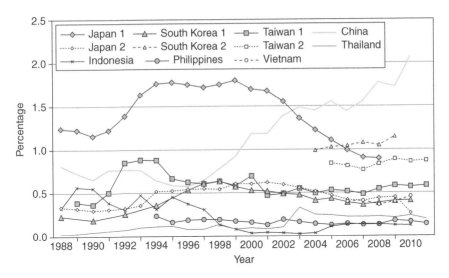

*Figure 1.1* Government environmental expenditure as a share of GDP (%) (sources: Badan Pusat Statistik (2002, 2007, 2011); Executive Yuan, R.O.C. (1996, 2003, 2012); Ministry of Environment, Japan (various years); Ministry of Environment, South Korea (various years); National Bureau of Statistics, China (various years); National Statistical Coordination Board of the Philippines (various years); National Statistical Office of Thailand (various years); and Chapter 5 in this book).

Notes
1 Japan 1, South Korea 1, Taiwan 1, and Thailand depict environmental expenditure of the central government only, while Japan 2, South Korea 2, Taiwan 2, and Vietnam include those of both the central and local governments.
2 South Korea includes budget for water supply because it is under the responsibility of the Ministry of Environment, while this is different in other countries.

while it accounted for 0.6 percent in South Korea and Taiwan, 0.4 percent in Indonesia and Vietnam, and 0.2 percent in Thailand and Philippines.

Third, East Asian governments have taken different combination of financing measures. There are several financing measures to increase state environmental expenditure, including: (a) increase environmental expenditure in the government budget, (b) mobilize domestic resources through tax and charges that may be earmarked for the environment, and (c) change the financing pattern of existing government institutions. Environmental fiscal mechanism in East Asian countries can be classified into several categories by the combination of these measures.

Japan and Taiwan increased their budget allocation for the environment in the general government budget and mobilized existing specialized governmental institutions. Both economies have a common feature in terms of how their governments integrated environmental considerations into industrial policy. Japan increased environmental expenditure in the budget, offered tax credits, and mobilized the existing postal savings to provide subsidy and/or subsidized loans to promote pollution control investment and industrial transformation as a package of stringent environmental regulations. Taiwan allocated the environmental budget for the same purpose, and offered tax credit, direct commercial banks to make loans at preferential terms of condition, while forcing state enterprises to make pollution control investments in tandem with implementing stringent environmental policies.

Thailand and Vietnam increased their budget allocation for the environment and created environmental funds. Thailand started to raise the government's environmental expenditure and created the national environmental fund in 1992. Partly due to bureaucratic segregation and competition over authority, and partly to decentralization, the share of government environmental expenditure has been stagnant, and the environmental fund has not worked well to finance environmental projects. The government has hesitated to implement wastewater treatment charges and to impose an environmental tax, regardless of the development of wastewater treatment plants at municipality level and promotion of cleaner production.

The Vietnamese government has taken a step further in terms of environmental fiscal reform. It created environmental protection funds both at the national level and in some large provinces, which are financed by government budget and environmental fines, and earmarked for soft loans and credit guarantees for environmental projects. It imposed a wastewater charge in 2003 and environmental tax to a wider range of wastes in 2012, although it has yet to become effective enough to reduce pollution (Yoshizumi and Hai 2012).

Indonesia spent a smaller share on environment in the general budget and did not have any fund that was earmarked for environmental protection. This accrues in part to its abundant oil revenues, which enabled it to obtain economic gains without serious environmental pollution (Cribb 1998). The exception was an afforestation fund that charged forest concession fees and was earmarked for afforestation activities. However, it was abandoned in the late 1990s as the government had often diverted its fund for other purposes.

China has also employed a variety of environmental fiscal measures. At the outset, it charged an environmental levy at the local level and created a local environmental fund that was earmarked for industrial pollution control projects and capacity building of environmental protection bureaus (EPBs) at local governments. This was a response to local governments' unwillingness to allocate their budget expenditure for environmental purpose, even for hiring staff at the EPBs. It was not until in the late 1990s that central policy makers increased government environmental investment, including pollution abatement investment at state enterprises at the national level, recognizing clear benefits of environmental projects funded by international aid (Mori 2009). The 1994 tax reform that allocated larger tax revenue to the central government enabled it to devote a larger share of the budget to the environment. As economic transition proceeded, the central government shifted its focal area from industrial pollution control toward environmental infrastructure development and the sloping land program, in which farmers have been paid in exchange for reforestation efforts.

By contrast, South Korea imposed various kinds of environmental charges and fees. The government initially imposed these charges and fees to complement the effectiveness of environmental regulations. However, a relatively small budget size coupled with a strict balanced budget principle directed the government to seek new fiscal sources for new activities. This led to the new charges and fees, coupled with creation of earmarked funds. It imposed 20 kinds of charges and fees, including air and water surcharges, recycling charges, and even contribution for ecosystem protection. In 1997 it changed the aim from revenue source to market-based instruments with the expectation that they work more effectively for emission reduction.

### Factors that have generated differences

There are several factors behind these differences. First, income level, industrialization process, and the severity of industrial pollution vary. All the Northeast Asian economies—Japan, South Korea, Taiwan, and China—faced massive industrial pollution and made huge investments to tame the rising anti-pollution movements. Southeast Asian countries, namely Thailand, Indonesia, Philippines, and Vietnam have also suffered from industrial pollution, but they have not yet recognized it as a nationwide problem that should be urgently addressed. Although they placed priority on deforestation, they tended to issue logging bans or give concession for commercial plantation and/or reforestation without charging higher rates for logging concession.

Second, the institutional arrangement of local government is different. Northeast Asian countries have mobilized local government budgets to manage the environment under their jurisdictions. Japanese local governments have spent considerable shares of their budget for developing wastewater and solid waste management system, and much more than the central government in the 1990s, for the latter encouraged the former to spend it as a fiscal stimulus. South Korea and Taiwan have enhanced the environmental management capacity of local governments from the outset, thus showing the same trend, although the share of local governments' spending has been larger in South Korea.[1] In China, local governments and state enterprises have been mandated to make environmental investment for new projects, and in part financially backed by a pollution levy and urban development fee that were collected at the local level.[2] In Vietnam, local governments have spent more than five times as large amounts for the environment as the central government (Chapter 5 in this book).

The third factor is the seriousness of government commitment to environmental policy and budget allocation to the environment. The Chinese government set the target for the share of state environmental investment in GDP in the tenth and eleventh Five-Year Plans, and achieved it in both periods. The Vietnamese government declared to allocate 1 percent of government expenditure for the environment both at the central and local level and has been struggling to achieve it. None of the other East Asian countries has published such a mid-term fiscal plan.

Nonetheless, China and Vietnam cannot achieve these commitments without fiscal prospects. In fact, they have steadily increased government revenue–GDP ratio, implying these governments have enjoyed higher revenue growth than economic growth (Figure 1.2). This high revenue growth enables them to increase the government's environmental expenditure and investment without compromising growing needs of the economic and social sectors. By contrast, Malaysia and Thailand show decreasing trends, reflecting preferential tax measures, tax incentives, and reduction of corporate tax rate to attract foreign investment and to promote export as a vehicle of industrialization, as well as weak tax administration. The government has to convince economic and social ministries to cut their spending when increasing environmental expenditure.

International environmental aid might also have a bearing on the seriousness of the government commitment. International donors have provided US$117 billion of official development assistance (ODA) that has had the environment as its primary purpose and as an important secondary purpose during 1995–2005, allocating a share of 38 percent to East Asian countries (Mori 2009). This has brought mixed results. On the one hand, such support showed visible environmental benefits to recipient governments, which eventually reduced aid reliance, as exemplified in China (Mori 2008). On the other hand, this environment-centered aid might turn

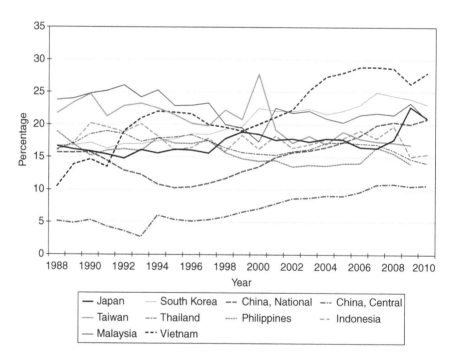

*Figure 1.2* Share of government revenue in GDP (source: same as Figure 1.1).

Note

Negative value means that the country is a net exporter.

out to be a mere substitute of the government environmental expenditure, discouraging East Asian governments to mobilize their domestic financing resources. Thailand had received ODA that had the environment as its primary purpose in the 1990s, which amounted to 73 percent of government environmental expenditure. With the international recognition that Thailand has become rich enough to manage the environment by itself, it has spent less than 3 percent of the government budget since 2003. However, the country has not increased government environmental expenditure so much to compensate for the decrease in international aid. Indonesia had for long sought international aid in the Suharto era and allocated it for development projects, including environmental infrastructure projects. Even after overcoming the confusion that was caused by the economic crisis, it maintains aid dependence (Table 1.1).

Finally, some countries are in the favorable position to win over public opinion and obtain society's acceptance for charges and fees. Many Japanese people, for instance, were pleased to pay wastewater treatment charge, acknowledging visible benefits to them: residents would replace traditional, unsanitary toilets to modern, clean ones once they paid charges and connected to the public sewerage system. Industrial users could save costs by discharging their wastewater to the public sewerage system, even when paying charges. In Southeast Asia, by contrast, residents and factories have often installed simplified treatment plants and regard treatment at public sewerage system as redundant. They do not want to pay charges to discharge wastewater even if their treatment plants are not sufficient to clean up wastewater.

### Limitation of domestic pressures

Despite domestic driving forces, most East Asian nations have insufficiently developed environmental policies and fiscal mechanism. Governments have placed priority on economic growth as a centripetal force for keeping their nation states together and for maintaining their legitimacy. East Asian authoritarian states have given the highest priority to such interests as the unity of the nation, the national prestige, and economic growth, while often sacrificing the interests of individuals, families, local communities, and ethnic minorities (Suehiro 1998). They adopted authoritarian regimes as a crisis management mechanism during the Cold War: they needed to concentrate power to suppress violent revolutions as well as the spread of communist revolutions. However, citizens got increasingly frustrated about the concentration of state power and political repression, and vitalized democratic movements. To keep their legitimacy, authoritarian governments have implemented growth-oriented policies to distribute material benefits to people in a visible manner, which resulted in the "growth first, cleanup later" strategy. Implementation of environmental policies and development of environmental fiscal mechanism is limited to the extent that they can tame environmental movements and keep economic growth intact.

Even if the regime is shifted to democracy, governments are not willing to implement unpopular policies that have significant distributional implications.

*Table 1.1* Ratio of environmental ODA to government environmental expenditure (%)

| | 1994 | 1995 | 1996 | 1997 | 1998 | 1999 | 2000 | 2001 | 2002 | 2003 | 2004 | 2005 | 2006 | 2007 | 2008 |
|---|---|---|---|---|---|---|---|---|---|---|---|---|---|---|---|
| China | 1.4 | 1.6 | 12.4 | 7.2 | 6.5 | 0.8 | 9.0 | 4.6 | 3.1 | 3.9 | 1.6 | 2.5 | 1.7 | 2.3 | 0.6 |
| Thailand | 76.5 | 0.7 | 40.8 | 44.1 | 115.4 | 178.6 | 80.9 | 15.6 | 10.3 | 3.9 | 2.4 | 4.3 | 4.2 | 0.9 | 2.1 |
| Indonesia | 61.1 | 53.3 | 69.4 | 0.6 | 304.8 | 65.3 | 162.5 | 108.3 | 62.2 | 140.6 | 93.8 | 98.6 | 43.5 | 14.4 | 86.3 |
| Vietnam | – | – | – | – | – | – | – | – | – | – | – | – | 109.3 | 85.1 | 191.8 |

Source: author calculation based on *Statistical Yearbook* depicted in Figure 1.1 and OECD-DAC (2010).

Note
In this table, ODA that has environment as its primary purpose is chosen to be on the safe side: government environmental expenditure is often defined as the expenditure of natural resource and environmental sector. Also the amount of ODA is commitment basis rather than disbursement that is often smaller, for recipient governments decide budget allocation in reference with the donors' commitment amount.

They hesitate to implement wastewater charges even after having developed wastewater treatment systems. This is true of low-income nations where the majority of people are incapable of paying their bills. Meanwhile, governments in high-income countries do not always set the tariffs high enough to recover costs unless they can provide visible benefits to them. Democratization also increased political influence of business groups which often lacked environmental concerns while urging acceleration of development projects even if they may have adverse impacts on environment and society, despite its prominent role in pushing government to establish democratic institutions (Mori 2012).

Increases in government environmental expenditure and earmarked fund have not always improved the ambient quality. Governments have invested in environmental infrastructure, such as water and waste treatment systems, but state-run utilities are often criticized for poor quality and inefficiency in service provision. This led to the privatization and private sector participation that are said to enhance efficiency, increase investment on infrastructure, and save government expenditure. In China, local EPBs diverted the pollution levy to their capacity building, financing less to industrial pollution control. This led to the pollution levy reform in 2003 that implemented the specific levy on sulfur—the most severe pollutant at that time—and absorbed part of the pollution levies' revenue to the central government for investing in pollution control across the regions.

## Global energy price hike and global climate governance

### *Energy tax, charges, and pricing before the mid-2000s*

Rapid economic growth has increased energy dependency in several East Asian countries. Japan, South Korea, and Taiwan have depended by more than 80 percent of energy consumption on import, while Thailand had to import more than 40 percent of its energy. China, which used to be a coal and oil exporter by the early 1990s, became an oil and gas importer in the early 2000s (Figure 1.3). While the net import is relatively small—more or less 10 percent of total energy consumption—its amount rapidly increased to reach 275 million tons of oil equivalent (ktoe) in 2009. The country is now the largest coal importer and second largest oil importer in the world.

Two oil crises in the 1970s alerted energy-poor East Asian countries to seek energy security. Japan, South Korea, and Taiwan are among the countries that have taken strong efforts toward achieving energy security. Rather than forging alliances, they competed with each other in establishing bilateral contracts with oil suppliers in the Middle East. This resulted in their weak bargaining power, which forced them to purchase oil at higher prices (Shim and Schreurs 2007).

Higher energy prices brought side effects: they encouraged energy-poor East Asian governments to improve energy efficiency and promote nuclear energy development as a way of diversifying energy sources as well as reducing air pollution from fossil fuel fired power plants. They implemented fuel tax and charges

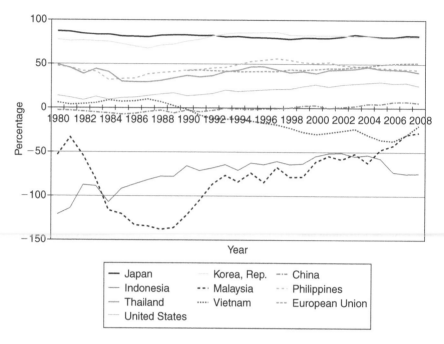

*Figure 1.3* Share of import energy in total consumption (source: World Bank 2011).

that were earmarked for developing nuclear power plants as well as getting social acceptance for development.

Energy-rich countries have capitalized ample oil and gas production to provide their constituents with energy at a subsidized price, which stimulated energy consumption. However, Malaysia has gradually reduced production and export due to declining capacity in the field. Indonesia increased exports of oil, oil products, and gas, but increased domestic consumption turned the country into a net oil importer in 2003.

China used to be an energy exporter, earning foreign exchange by exporting coal and oil during the 1980s, while often facing tradeoff among insufficient supply, increase in coal reliance and inefficient energy use, and serious air pollution. To overcome this tradeoff, it instigated the removal of energy subsidies, forced the closure of small and inefficient coal mines, and implemented a sulfur levy. But these measures resulted in energy shortages and frequent rotating blackouts. In response, the government increased the domestic coal price and abandoned differentiated pricing between electricity use and non-electricity use, and raised the coal tax, the resource compensation fee, and the resource exploitation royalty. But government strict control on the retail electricity price made it difficult for thermal power generation and transmission companies to transfer the increasing fuel costs, thus making them fall into financial deficit (Horii 2010).

### Impact of global energy price hike in the mid-2000s

The global energy price hike, coupled with recovering energy demand hit the government budget most in Indonesia. The fuel subsidy jumped up to 18.6 percent and 22.6 percent of government expenditure in 2005 and 2008, respectively (Table 1.2). To prevent the fiscal deficit from widening, the Indonesian government increased the fuel price by 125 percent in 2005 and 28.7 percent in 2008 on average, respectively. These oil subsidy cuts temporally reduced government expenditure, but faced a rebound effect. The government prepared another fuel price hike by 33 percent in 2012, but when faced with fierce citizens' protests, abandoned its plans at last.

Global energy price hikes also increased power production cost and subsidies due to high dependence of the State Electric Company (PLN) on oil. Coupled with growing electricity consumption in the 2000s, this caused power shortages, frequent blackouts, and interruption in the production process. In response, the government published the National Energy Policy in 2006 that set an industrial energy intensity reduction target, and charged the PLN with carrying out a crash program to build coal-fired power plants with a total capacity of 10,000 megawatts, with the goal of phasing out oil fuel fired power plants to reduce subsidies. After giving up its plans to raise the fuel price in 2012, the government shifted the focus on the hike of electricity tariffs.

Japan took contradictory policies. On the one hand, it implemented the fuel cost adjustment system so that electricity power companies could transfer changes in import energy prices and exchange rates quickly to consumers without government permission. On the other hand, parliament decided not to extend a "temporary" gasoline tax that was earmarked to develop road infrastructure, to mitigate the impact of energy price hike on retail gasoline price. The government passed the law to extend the tax at last, but adding the clause that it would not charge a gasoline surcharge when the price goes beyond the predetermined price.

China implemented an energy tax and pricing reform. It increased the transport fuel tax in exchange for abolishing toll fees and fuel excise in the road maintenance and management fee in 2008. To satisfy growing energy demand while reducing coal consumption, it mandated transmission companies to purchase renewable electricity at a premium price. However, slow progress in electricity sector reform, especially in retail price reform prohibits renewable

*Table 1.2* Share of environmental expenditure and oil subsidy in the government expenditure in Indonesia, 2005–2010 (%)

|  | 2005 | 2006 | 2007 | 2008 | 2009 | 2010 |
| --- | --- | --- | --- | --- | --- | --- |
| Environmnetal expenditure | 0.61 | 0.66 | 0.72 | 0.68 | 0.69 | 0.74 |
| Oil subsidy | 18.8 | 9.6 | 15.4 | 22.6 | 9.2 | 13.1 |

Source: author compilation based on Badan Pusat Statistik (2011).

electricity from being connected to the transmission line. This threatens energy security in China, which is regarded as critical to sustain economic growth and prevent social unrest. This situation has driven the Chinese government into energy saving and import from all over the world.

South Korea implemented a transport tax with the transport, energy, and environmental tax in 2007 and continued its educational and local drive surtax. But these charges and taxes have not effectively worked for demand management, because most energy-intensive sectors are given tax reductions and exemptions. In addition, the government has spent a large part of the revenue to increase energy and transport infrastructure that would induce future demand. In the 2007 Basic National Energy Plan, it assumed to increase nuclear power and renewable energy to meet the increasing demand, and to implement demand-side management for energy efficiency.

When both currency devaluation and global energy price hike induced the rise of retail energy prices, the Korean government took countermeasures to maintain its industrial competitiveness and to avoid inflation. It instigated an electricity price reform in 2008 to set the night time price lower for industrial use—even lower than imported oil price. This pricing policy has increased energy consumption and made the Korean Electric Power Company (KEPCO) fall into deficit in 2009 and a major blackout in the Seoul Metropolitan Area in September 2011. The Korean government eventually admitted an electricity price hike for residential use by 4.9 percent in August 2012 and by another 4.5 percent in December 2012, but continues to hesitate allowing a price hike for industrial use.

Thailand reintroduced diesel subsidies in 2010 to tackle the rising international oil price and to keep retail prices below 30 THB per liter. Yet, facing the depletion of the Oil Fund reserve, the Thai government replaced it with a reduction of the excise tax, at the expense of government tax revenue.

### *Impact of emerging global climate governance*

The Copenhagen Accord of December 2009 was a turning point for East Asian countries to make serious commitment to greenhouse gas (GHG) emissions reduction. Although no binding reduction target was globally agreed upon, South Korea, China, Malaysia, and Indonesia submitted their Nationally Appropriate Mitigation Actions (NAMA) to the UN Framework Convention on Climate Change (UNFCCC).

To achieve their mid-term GHG emissions reduction target, they have advanced the existing policies as well as implementing new ones. The Chinese government set ambitious targets on the reduction of energy and carbon intensity, and increase in the share of non-fossil fuel in the twelfth Five-Year Plan (2011–2015). It imposed further responsibility on energy saving to local government leaders, implemented pilot emission-trading schemes (ETS) at several provinces, and designated cities to expand it nationwide in near future. There was an internal debate on carbon tax, but the government only instigated pilot projects in two provinces that upgraded the existing pollution levy while not containing carbon emission in the base.[3] Still, this design raises controversy over authority, for it deprives local EPBs of authority for collecting and spending pollution levy and may be the only environmental fiscal source for environmental protection (Ge *et al.* 2011).

South Korea enacted the Framework Act on Low Carbon Green Growth in 2009, followed by the mid-term targets on renewable energy and energy intensity, and replacement of feed-in-tariff to renewable portfolio standards to mandate power companies to generate a specified fraction of the power from renewable sources. However, the government continues to tout nuclear as the primary power source for Korea's future *green* growth. Korea became the third largest generator of nuclear power in total domestic electricity generation in 2011, and plans to increase its share from 36 percent of total 2007 power generation to 59 percent in 2030, thereby absorbing the largest share of the budget for green technology development (35.9 percent in 2009) (Yun *et al.* 2011).

Japan, to attain its 25 percent reduction target by 2020, prepared legislation for an emission trading system, a carbon-energy tax, and a feed-in-tariff for renewable energy, while attempting to increase nuclear power as a "cost-effective" measure in the mid-term energy plan in 2010. Faced with fierce opposition from industrial associations, the Japanese government implemented a carbon-energy tax in a small scale and gave up its plans of implementing an emission trading system. However, the Fukushima nuclear accident disabled the government to operate nuclear power at pre-disaster levels, let alone to commission new ones. It expanded a feed-in-tariff to a variety of renewable energy in 2012, with the aim of substituting nuclear power rather than reducing carbon emissions.

Indonesia announced the launch of the second crash program in 2010 to increase power generation capacity. Although the new capacity is the same as in the first program, it aimed to attain the GHG emissions reduction target: it would build renewable power generation more than half of total new capacity while coal-fired plants only one-third (JETRO, 2011). To promote renewable electricity, the government implemented several tax incentives.

### *Why have countries taken different policy responses?*

From the above description, we can find at least two factors that have led East Asian countries to take different responses to the global energy price hikes and global climate governance.

First one is energy endowment. Energy-poor countries of South Korea, Thailand and Japan had already imposed energy taxes and charges with the aim of gaining energy security: these taxes and charges are often earmarked for developing alternative energy, exploiting new oil and gas field for export, and encouraging energy efficiency. As long as these funds work effectively for securing cheap energy, the government can easily get social acceptance. By contrast, energy-rich countries have long provided subsidies to enjoy cheap energy, during accelerated domestic consumption.

Second factor is the balance between government budget and social resistance to energy price hike. When faced with global energy price in the 2000s, South Korea and Thailand prevented an energy price hike at the expense of government budget. Japan removed the fuel tax "temporally", while enabling power companies to pass on energy price hike. China implemented a transport fuel tax, while it was cautious about a retail price hike. Large fiscal deficit forced Indonesia to cut fuel subsidies, while fierce social protests prevented them in 2012.

The third one is government initiative and social acceptance to nuclear power generation, which enables countries to reduce GHG emissions and to avoid losing industrial competitiveness even if they implement an emission trading system. China, South Korea, and Vietnam keep their momentum while Japan lost after the Fukushima nuclear accident in 2011. This may discourage them from implementing a carbon-energy tax that may have impacts on distribution and industrial competitiveness that East Asian countries have highest concerns to maintain their development model.

## Summary and outlook

We can summarize our findings as follows.

First, East Asian countries have developed environmental fiscal mechanism in tandem with the development of environmental institutions and policies, sought fiscal sources for enhancing government environmental capacity and/or environmental investment, with the exception of Indonesia which preferred an information-based policy instrument. Yet, per capita income, changes in government fiscal revenue and social environmental awareness brought about the difference in public acceptance of environmental charges and fees, which pushed individual nations to develop different environmental fiscal mechanisms.

Second, the recent global energy price hike and emerging global climate governance have added momentum for carbon-energy fiscal reform, while the extent varies, reflecting different energy endowments and cost and benefit profiles that global climate institution offer. While recent development of global climate governance pushes East Asian countries to implement ETS and promote renewable and nuclear energy, it may not always provide favorable conditions for furthering carbon-energy fiscal reform.

## Notes

1 This difference accrues partly to the fact that local governments have their own environmental revenue sources in South Korea while this form of local fiscal autonomy does not exist in Taiwan.
2 Their activities are reflected in the statistics that count on environmental investment of the central and local governments and state enterprises, while not counting on administrative and management expenditure.
3 Hu Yang, "Environmental tax plan passed by ministries," *China Daily* December 7, 2010.

# References

Badan Pusat Statistik (various years), *Statistical Yearbook of Indonesia*, Jakarta: BPS.

Ban, H. and Fujikawa, K. (2013), "Cost and benefits of regional market-based environmental policy: East Asian carbon market", in Mori, A. (ed.), *Environmental Governance for Sustainable Development: An East Asian Perspective*, Tokyo: United Nations University Press, 257–70.

CCICED (2005), *Environmental and Natural Resources Pricing and Taxation*, CCICED Task Force Report on Environmental and Natural Resources Pricing and Taxation, Beijing. Online, available at: www.sfu.ca/international-development/cciced/pdf/2005_TFonPricingandTaxation.pdf (last accessed: January 14, 2013).

CCICED (2009), *China's Pathway Towards a Low Carbon Economy*, CCICED Policy Research Report, Beijing. Online, available at: www.sfu.ca/international-development/cciced/pdf/2009_TFonLowCarbonEconomy.pdf (last accessed: January 14, 2013).

Cribb, R. (1998), "Environmental policy and politics in Indonesia", in Desai, U. (ed.), *Ecological Policy and Politics in Developing Countries*, Albany, NY: State University of New York Press, 65–85.

Executive Yuan, R.O.C. (various years) *Statistical Yearbook of the Republic of China*, Taipei: Directorate General of Budget, Accounting and Statistics (DGBAS) of Executive Yuan.

Ge, C., Gao, S., Ren, Y., Sun, G., and Long, F. (2011), "Study on the design of wastewater environmental tax in China: from wastewater pollutant discharge fee to environmental tax", in Kreiser, L., Sirisom, J., Ashiabor, H., and Milne, J.E. (eds), *Environmental Taxation in China and Asia-Pacific: Achieving Environmental Sustainability through Fiscal Policy*, Cheltenham: Edward Elgar, 66–83.

Horii, N. (2010), "Background of resource and electricity bottlenecks and future supply constraints", in Horii, N. (ed.), *China's Sustainable Growth: Can China Overcome Resource and Environmental Constraints?* Chiba, Japan: IDE-JETRO, 23–56 (in Japanese).

Inui, T., Morishima, A., and Imura, H. (2005), "Japan's environmental policy: financial mechanisms", in Imura, H. and Schreurs, M.A. (eds), *Environmental Policy in Japan*, Cheltenham: Edward Elgar, 249–65.

JETRO (2011), "Renewable Energy Policy in the Asia Pacific", Survey Report. Online, available at: www.jetro.go.jp/jfile/report/07000695/asia_pacific_reuse_energy.pdf (last accessed: February 10, 2013) (in Japanese).

Lee, S.C. (ed.) (2010), *Environmental Charges in East Asia: Factors and Barriers for Institutional Evolution*, Kyoto, Japan: Showado (in Japanese).

Liu, X., Ogisu, K., Suk, S., and Shishime, T. (2011), "Carbon tax policy progress in North-east Asia", in Kreiser, L., Sirisom, J., Ashiabor, H., and Milne, J.E. (eds), *Environmental Taxation in China and Asia-Pacific: Achieving Environmental Sustainability through Fiscal Policy*, Cheltenham: Edward Elgar, 103–18.

Ministry of Environment, Japan (1971–2003), *Quality of the Environment in Japan*; (2004–2006), *Annual Report on the Environment in Japan*; (2007–2012), *Annual Report on the Environment, the Sound Material-Cycle Society and the Biodiversity in Japan*.

Ministry of Environment, South Korea (various years), *Environmental Statistical Yearbook*.

Mori, A. (2008), "Impacts of Japanese ODA on the development of China's environmental policy and institutions", in Mori, A., Ueta, K., and Yamamoto H. (eds), *Environmental Policy in China*, Kyoto, Japan: Kyoto University Press, 275–303 (in Japanese).

Mori, A. (2009), *Environmental Aid: Logic, Strategy and Evaluation to Advance Sustainable Development*, Tokyo: Yuuhikaku (in Japanese).

Mori, A. (2012), "Introduction", in Mori, A. (ed.), *Democratization, Decentralization and Environmental Governance in Asia*, Kyoto, Japan: Kyoto University Press, 1–25.

Mori, A., Lee, S.C., and Ueta, K. (2005), "Economic incentives to promote compliance: Japan's environmental soft loan program", in Bianchi, A., Cruz, W., and Nakamura, M. (eds), *Local Approaches to Environmental Compliance: Japanese Case Studies and Lessons for Developing Countries*, Washington, DC: World Bank Institute, 89–100.

National Bureau of Statistics, China (various years), *China Statistical Yearbook on Environment*, Beijing: China Statistical Press.

National Statistical Office of Thailand (various years), *Statistical Yearbook of Thailand*, Bankok: National Statistical Office of Thailand.

National Statistical Coordination Board of the Philippines (various years), *Philippines Statistical Yearbook*, Manila: National Statistical Coordination Board of the Philippines.

OECD (1999), *Environmental Taxes: Recent Developments in China and OECD Countries*, Paris: OECD.

OECD-DAC (2010), *International Development Statistics 2010 CD-ROM*, Paris: OECD.

Park, S.J., Yamazaki, M., and Takeda, S. (2012), *Low Carbon Green Growth Roadmap for Asia and the Pacific: Major Findings and Policy Implications from a Multi-Regional Economic Simulation Analysis*, Bangkok: UN ESCAP.

Shim, S. and Schreurs, M.A. (2007), "Energy security and regional stability in Northeast Asia", in Hyun, I.T. and Schreurs, M.A. (eds), *The Environmental Dimension of Asian Security: Conflict and Cooperation over Energy, Resource and Pollution*, Washington, DC: United States Institute of Peace, 41–61.

Suehiro, A. (1998), Developmentalism in developing countries. In Institute of Social Science, University of Tokyo (ed.), *The 20th-Century Global System 4: Developmentalism*, Tokyo: University of Tokyo Press, 13–46 (in Japanese).

Ueta, K., Lee, S.C., Chen, L., and Jin, H. (2009), "Environmental policy and environmental public finance: toward comparative analysis on environmental fiscal mechanism", in Mori, A. (ed.), *Economic Development and Environmental Policy in East Asia*, Kyoto, Japan: Minerva Shobo, 148–75 (in Japanese).

Wang, J., Ge, C., and Yang, J. (eds) (2003), *Environmental Financing Strategy*, Beijing: China Environmental Science Press (in Chinese).

Wang, J., Ge, C., Gao, S., and Ren, Y. (2011), "Policy design of environmental tax in China", in Kreiser, L., Sirisom, J., Ashiabor, H., and Milne, J.E. (eds), *Environmental Taxation in China and Asia-Pacific: Achieving Environmental Sustainability through Fiscal Policy*, Cheltenham: Edward Elgar, 18–34.

Wang, W. (2011), "Greening the dragon: energy tax policy in China", in Kreiser, L., Sirisom, J., Ashiabor, H. and Milne, J.E. (eds), *Environmental Taxation in China and Asia-Pacific: Achieving Environmental Sustainability through Fiscal Policy*, Cheltenham: Edward Elgar, 3–17.

World Bank (2011), *World Development Indicator 2011*, Washington, DC: World Bank.

Yoshizumi, M. and Hai, H. (2012), "Vietnam: progress and challenges in environmental policy", in Mori, A. (ed.), *Environmental Policy in East Asia*, Kyoto, Japan: Showado, 160–75 (in Japanese).

Yun, S.J., Cho, M.R., and von Hippel, D. (2011), "The current status of green growth in Korea: energy and urban security", *Asia-Pacific Journal* 9 (44), October 31.

# 2 China's fiscal expenditure for environmental protection

*Zhong Ma and Jian Wu*

Based on the overview of China's fiscal policies from 1993 to 2009, this chapter analyzed China's fiscal expenditure for environmental protection in three aspects: Account for Environmental Protection, environmental taxation system and ecological compensation. As education and environmental protection are both public services, this chapter made a comparative analysis of fiscal expenditure for education and environmental protection. It concludes that it is rather urgent to increase the state's fiscal expenditure for environmental protection.

## Overview of China's fiscal policies: 1993–2009

In the early 1990s, China established a more market-oriented economy, and the market was acknowledged to play the fundamental role in allocating resources. Fiscal policies are the instruments utilized by the government for macro-economic regulation. Since 1993, the Chinese government has been implementing a tight fiscal policy to regulate the economy and provide public services.

### Tight fiscal policy: 1993–1997

In 1992, China's economy entered a new round of rapid growth. In 1993, the economy trended toward overheating and the growth rate of total investments in fixed assets reached 61.78 percent. The growth rate of total retail sales of social consumer goods was 29.81 percent. The Commodity Retail Price Index increased by 21.7 percent. In June 1993, the Party's Central Committee and the State Council issued "The Opinions on the Current Economic Situation and Strengthening the Macro-regulation", and marked the implementation of tight fiscal policy. The policy requested that in order to maintain steady economic growth and prevent inflation, expenditure reduction and structural adjustment were to be implemented. The country's key construction projects and the infrastructure for public service were considered as investment priorities, including roads, railways, farmland infrastructure, water conservation facilities, and environmental protection projects.

The tight fiscal policy included the following measures:

1 From January 1, 1994, the tax-share financial management system has been formally implemented, which aims to increase the proportion of the fiscal revenue to GDP and the central fiscal revenue to the national fiscal revenue in order to provide financial guarantees for the government's macro-regulation.

2 The *Budget Law of the People's Republic of China* that was effective on January 1, 1995, stipulates the following aspects:

   • the public budget of the central government shall not run into a deficit;
   • partial funds for construction investment indispensable to the central budget may be raised in form of domestic and foreign loans;
   • the local budgets at various levels shall not run into a deficit; and
   • local governments cannot issue local government bonds.

   Sequentially, the national fiscal deficit ratio (the ratio of the fiscal deficit to GDP) was reduced from 1.19 percent in 1994 to 0.74 percent in 1997.

3 Control the growth rate of fiscal expenditure and adjust fiscal expenditure structure toward the weaker sectors of the economy. Annual growth rate of fiscal expenditure fell from 24.8 percent in 1994 to 16.3 percent in 1997. During the period of 1994–1996 the nation's fiscal expenditures in agriculture increased by 16.7 percent, which was 4 percent higher than the 1991–1993 average growth. National fiscal expenditure in environmental protection was also increased. In 1996 and 1997, the total investment in environmental protection reached RMB91.07 billion (Xie, 2008).

4 Strengthen tax collection and administration, control tax reduction or exemption through various approaches including administering tax by law, rectifying weaknesses in tax collection and administration, and finishing the issuance of treasury bills before the deadlines.

5 Remove and compress infrastructure projects, and control the excessive growth of investment in fixed assets.

6 Strictly control the budget of the governments and restrain the excessive growth of consumption.

After four years of regulation by the tight fiscal policy, the growth rates of total investment in fixed assets and total retail sales of social consumer goods were reduced (Figure 2.1). Per capita resident income continued to grow and the fiscal revenue rose from RMB434.90 billion in 1993 to RMB865.11 billion in 1997, i.e., nearly doubled over four years.

### Expanding fiscal policy: 1998–2004

July 1997 marked the start of the Asian financial crisis. The growth rate of GDP in the first half of 1998 dropped to 7 percent, and the economy showed signs of gradual deflation and slow growth. This economic situation pushed the transition from a tight fiscal policy to an expanding fiscal policy. In July 1998, the State

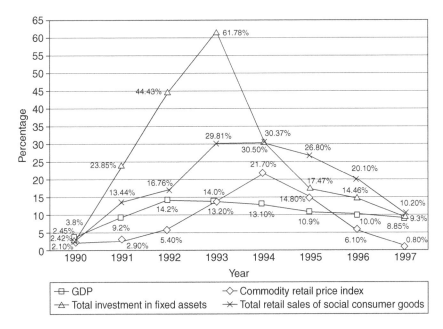

*Figure 2.1* Annual growth rate of GDP, commodity retail price index, total investment in fixed assets, and total retail sales of social consumer goods, 1990–1997 (source: Statistical Database of China Economic Information Network in National Bureau of Statistics of China 2010).

Council decided to implement the expanding fiscal policy to stimulate demand, i.e. to increase expenditure and investment, stimulate consumption, expand exports, and promote stable growth of the macro economy by selling long-term construction treasure bonds without implementing tax cuts.

The expanding fiscal policy included the following measures:

1   A total of RMB910 billion long-term construction treasury bonds were released during the years 1998–2004. At the end of 2004, RMB864.3 billion had been invested in agriculture, forestry, water conservation, ecological construction, transportation and communication infrastructure, urban infrastructure, technological innovation and industrial upgrading, rural grids, and education, culture, health, tourism, infrastructure construction, and environmental protection investment (Figure 2.2).

2   Increase public investment. The national budget deficit amounted to RMB314.951 billion in 2002, amounting to a fiscal deficit ratio of 2.62 percent. In 2004 this ratio fell to 1.31 percent (Figure 2.3).

3   Adjust tax policy to support export, promote foreign investment, and reduce the burden on enterprises.

4   Adjust income allocation policy to foster and expand consumption demand. In the period of 1999–2003, the government increased the wage standard by

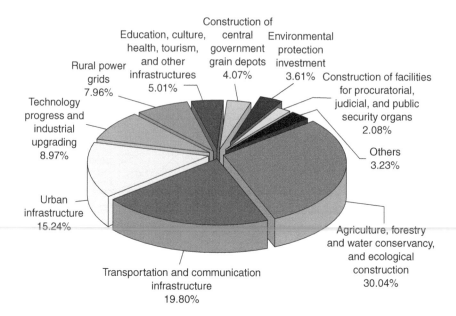

*Figure 2.2* Allocation of long term treasury bonds, 1998–2004 (source: Statistical Database of China Economic Information Network in National Bureau of Statistics of China 2010).

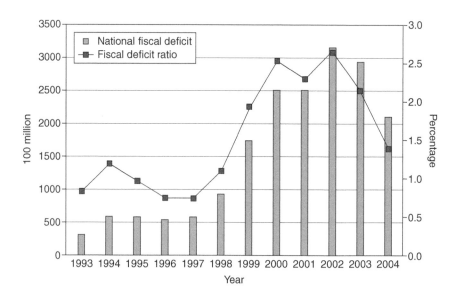

*Figure 2.3* Annual national fiscal deficits and fiscal deficit ratio, 1993–2004 (source: Statistical Database of China Economic Information Network in National Bureau
of Statistics of China 2010).

four times for government staff. Fiscal expenditure on social security accounted for 12.4 percent of the total fiscal expenditure in 2004, which was 7.2 percent in 1998.

5    Improve non-tax revenue policy to strengthen the financial management system. During the years of 1994–2004, the government canceled approximately 2000 charge programs and reduced the rates of nearly 500 fee programs. The government increased transfer payments to the middle and western regions and to rural areas. During 2002–2004 transfer payments for rural tax and fee reform amounted to RMB107.3 billion.

6    Support the adjustment of economic structure and promote the reform of state-owned enterprises.

The expanding fiscal policy helped China enter into a new growth cycle, in which GDP, investment demand, and export showed a momentum of continually rapid growth (Figure 2.4). Per capita net income of rural households increased from RMB2090 in 1997 to RMB2936 in 2004. Per capita disposable income of urban residents rose from RMB5160 in 1997 to RMB9422 in 2004.

By 2004, the fiscal revenues reached RMB2.64 trillion, i.e., doubled compared to that of 1997. The ratio of fiscal revenue to GDP rose from 10.95 percent in 1997 to 16.51 percent in 2004. The ratio of central government revenue to national fiscal revenue increased from 48.9 percent in 1997 to 54.9 percent in 2004. Government fiscal resources, especially the rapid growth of central government fiscal resources created favorable conditions for the establishment of the public fiscal framework.

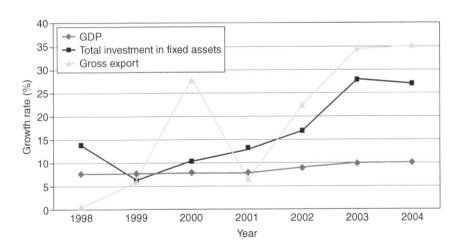

*Figure 2.4* Annual growth rates of GDP, total investment in fixed assets, and gross export, 1998–2004 (sources: Statistical Database of China Economic Information Network in National Bureau of Statistics of China 2010).

*Neutral fiscal policy: 2005–2008*

Since the second half of 2003, China's economy has suffered from the following problems:

- investment growth has been excessive and inflationary pressure has increased;
- scarcity of resources and environmental quality has become apparent;
- resource allocation in the areas of social development was inadequate, especially in agriculture, public education, public health, and social security; and
- structural and institutional problems were becoming increasingly prominent.

The public fiscal system was identified as the pattern of China's fiscal reform, and fiscal policy stepped into a new round of strategic adjustment again. In December 2004, China's Central Economic Conference promulgated the neutral fiscal policy which was implemented in 2005. The neutral fiscal policy was to "control deficit, adjust structure, promote reform, increase revenue and reduce expenditure". The implementation of neutral fiscal policy went along with a series of fiscal reforms, including:

1   Reduce the fiscal deficit and control the issuance of treasury bonds. The ratio of the national fiscal deficit to GDP fell from 1.31 percent in 2004 to 0.07 percent in 2008. The scale of the long-term construction treasury bonds was RMB80 billion, RMB60 billion, RMB50 billion, and RMB30 billion subsequently from 2005 to 2008.
2   Adjust and optimize the structure of financial expenditure. Essentially the total amount of financial expenditure was not changed fundamentally. The fiscal expenditure structure and treasury bonds project orientation were appropriately adjusted, that gradually exited from excessive investment sectors, and supported weak investment sectors, including agriculture, employment and social security, environmental protection and ecological construction, energy saving, public health, education, and science and technology.
3   In 2006, agricultural tax was rescinded nationwide. In order to establish a sound public fiscal system, the so-called Classification Reform Program on Government Revenue and Expenditure that was developed by the Ministry of Finance in February 2006 adopted environmental protection as a classification subject in the national fiscal account. This is the first time that the national fiscal accounting system has included provisions for environmental protection, which was formally implemented on January 1, 2007 (Figure 2.5).

In the period of the neutral fiscal policy, China's economy maintained rapid growth. GDP growth rate in 2007 was 13 percent which was the highest in the preceding decade. Per capita net income of rural households increased from RMB2936 in 2004 to RMB4140 in 2007. Per capita disposable income of urban residents was enhanced from RMB9422 in 2004 to RMB13,786 in 2007. In 2007, the national fiscal revenue reached RMB5.13 trillion, nearly double the amount in 2004.

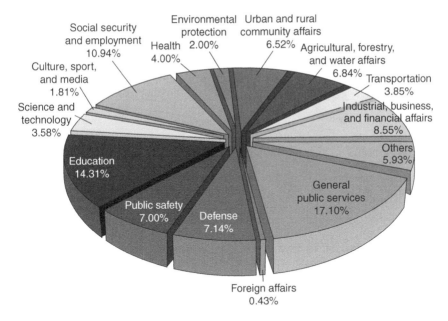

*Figure 2.5* National fiscal expenditure structure, 2007 (source: Ministry of Finance 2007).

## *Expanding fiscal policy: 2008 to present*

In the context of the eruption of the 2008 global financial crisis, China's economic growth slowed down significantly, and GDP growth rate in the third and fourth quarter of 2008 was 9 percent and 6.8 percent. In order to expand domestic demand and promote stable economic growth, the State Council Executive Session that was held on November 5, 2008, claimed the implementation of the expanding fiscal policy. In December 2008, China's Central Economic Conference clearly demonstrated the main content of the expanding fiscal policy: substantially increase public expenditure, guarantee the expenditure in key areas and key construction, support the post-earthquake recovery and reconstruction, implement structural tax cuts, and optimize expenditure structure. This meant that the expanding fiscal policy includes two expansionary approaches of revenue reduction and expenditure increase.

The expanding fiscal policy included the following specific measures:

1   Expand the deficit and issue treasury bonds to increase public investment expenditure. The central government launched an added RMB4 trillion stimulus package for two years that started from the fourth quarter of 2008. The central government invested RMB1.18 trillion (accounting for 29.5 percent) to promote local funding RMB2.82 trillion (accounting for 70.5 percent) (Figure 2.6). The promulgated national budget deficit in the two years of 2009–2010 was RMB2 trillion.

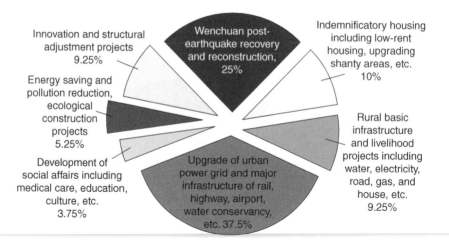

*Figure 2.6* Key investment areas of the four trillion RMB stimulus package (source: National Development and Reform Commission 2009).

2   Optimize fiscal expenditure structure. The government tended to strictly control the general expenditure budget, and continually increased the expenditure in agriculture and rural development, livelihood improvement, independent innovation, energy saving, and support of less developed areas. A substantial increase in expenditure was evident in five categories in 2011, namely education, health insurance of farmers and urban residents, medical institutions, social security investment, indemnificatory housing, and water conservation measures.

3   Implement a series of structural tax cuts. In addition to continue the implementation of 2008's tax cuts, the government has implemented a series of tax reduction and exemption policies, including a nationwide implementation of the consumption-type VAT, the implementation of an oil tax reform, cancellation and suspension of 100 administrative fees, etc.

4   Strengthen the government debt management and enhance internal and external binding to effectively mitigate and defuse potential financial risks.

In 2009, China's GDP reached RMB33.5 trillion, with a growth rate of 8.7 percent. Per capita disposable income of urban residents was RMB17,175 and per capita net income of rural residents was RMB5153, with the actual growth rate of 9.8 percent and 8.5 percent respectively. Fiscal revenue in 2009 was RMB6.85 trillion, with an increase of 11.7 percent compared to that of 2008, which was the lowest increase in 15 years.

## China's fiscal and taxation reform

In December 1993, the State Council released the decision on Reform of Fiscal and Taxation System, which came into effect on January 1, 1994. The core of the reform was to set the separate local and central taxation systems, including four components:

1 Defining the fiscal expenditure scopes of the governments in different levels based on the administrative rights of central and local governments. The central government is taking care of the expenditures for the country's security, foreign affairs, and institutional operation, while the local governments are taking care of the expenditures of the corresponding government operation, social, and economic development.
2 Dividing the revenues of central and local governments by tax categories based on the integration of administrative and fiscal rights and establishing two tax administrative systems for local and central respectively.
3 Gradually implementing the fiscal return and transfer system from central to local governments. The returned tax annually increased at the basis of 1993.
4 Establishing and defining the grading budgetary system and setting up the constraints on the government budgets in various levels.

Since starting the fiscal reform, the revenues of both central and local governments have been growing rapidly. In 2009 the total revenue was RMB6.9 billion with central government's revenues accounting for 3.6 billion, while the local governments' revenue was 3.3 billion, which signify increases by 11.02 and 10.90 times respectively compared to the 1995 numbers (Figure 2.7).

During the period of 1995 to 2009, the average annual growth rates for central and local governments' revenues were 18.4 percent and 19.44 percent respectively, which is much higher than the average annual growth rate of GDP (around 10 percent) during the same period (Figure 2.8).

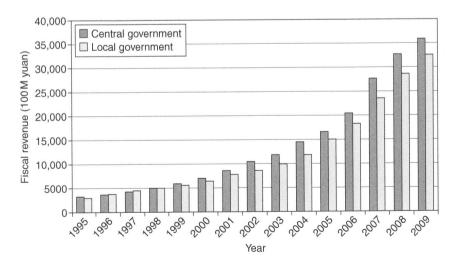

*Figure 2.7* Central and local government revenues in China, 1995–2009 (source: National Bureau of Statistics of China 2010).

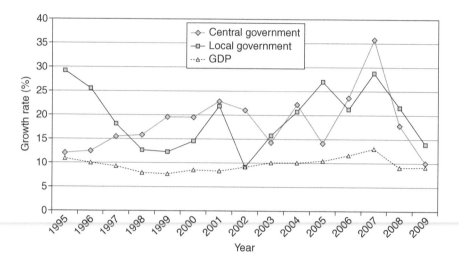

*Figure 2.8* Central and local governments' revenue and GDP growth rate in China, 1995–2009 (source: National Bureau of Statistics of China 2010).

## Account for Environmental Protection

The State Council issued the *Decision on Implementing the Scientific Concept of Development and stepping up Environmental Protection* in December 2005, indicating that governments at all levels should include environmental protection investment in fiscal expenditure and increase year by year, so as to improve the environmental investment mechanism. In addition, investments in pollution prevention, ecological protection, *demonstration sites* of environmental protection, and capacity building for environmental monitoring and management must be enlarged. In February 2006, the Account for Environmental Protection (shorten as Subject 211) was established as one of the subject items of state fiscal budget by the Ministry of Finance (MOF). Therefore, expenditure for environmental protection has been included in fiscal budget since the reform carried out on January 1, 2007.

### *Contents of the Account for Environmental Protection*

The Account for Environmental Protection includes 14 terms as follows:

* environmental protection administration expenditure;
* environmental monitoring and supervisory expenditure;
* pollution control expenditure;
* ecological protection expenditure;
* natural forest protection program expenditure;
* *reforesting cultivated land project* expenditure;

- sandy desertification management expenditure;
- replacing cropland with pasture expenditure;
- replacing cropland of cultivated grassland expenditure;
- energy saving expenditure;
- pollution emission reduction expenditure;
- renewable energy expenditure;
- comprehensive utilization of resources expenditure;
- energy management expenditure.

In short, they can be summarized as three categories: pollution control, ecological protection, and energy.

### Overview of Subject 211 expenditure, 2007–2009

Regarding the amount, fiscal expenditure on Subject 211 for environmental protection increased dramatically from 2007–2009, i.e., from RMB99.6 billion (2007) to RMB145.1 billion (2008) and to RMB193.4 billion (2009).

With respect to composition, according to fiscal expenditure for environmental protection in the national fiscal expenditure balance sheet, reforesting cultivated land project had the largest share, increasing from 21.14 percent in 2008 to 22.66 percent in 2009, and energy saving followed by a slight decrease from 10.72 percent in 2008 to 10.18 percent in 2009. Replacing cropland with pasture had the smallest share, which was 1.35 percent in 2008 and 1.89 percent in 2009. Moreover, it is worth noting that the share of renewable energy was relatively low, at only about 3 percent (Figure 2.9).

With respect to the following three categories: pollution control, ecological protection, and energy, according to fiscal expenditure for environmental

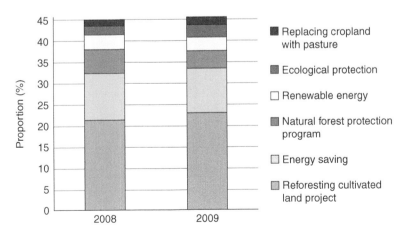

*Figure 2.9* Composition of Subject 211, 2008–2009 (source: Ministry of Finance 2008–2009).

protection in the national fiscal expenditure balance sheet, pollution control had the largest share (such as pollution control technology), accounting for over 50 percent, and ecological protection (such as *reforesting cultivated land*) followed with a slight increase from 30.43 percent in 2008 to 31.5 percent in 2009. In addition, energy only took up 13.80 percent in 2008 and 13.24 percent in 2009 (Figure 2.10).

In 2009 the country's total fiscal expenditure was RMB7.63 trillion, increased by 21.9 percent from 2008, and fiscal expenditure for environmental protection increased dramatically. However, the expenditure for environment is much less for than rural development, education, medical and health, and scientific and technology development (Figure 2.11).

### Conclusions

Subject 211 for environmental protection includes 14 items, and can be summarized as three categories: pollution control, ecological protection, and energy.

Pollution control accounts for the largest share (over 50 percent). However, it is worth noting that some part of pollution control expenditure is collected from the pollution charges. The proportion of pollution control expenditure in the total fiscal expenditure is only slightly higher than 1 percent. For example, in 2009, pollution control expenditure was approximately RMB106.887 billion, RMB17.26 billion of which was from pollution charges. Pollution control expenditure accounted for only 1.17 percent of the fiscal funds. Besides, the expenditures for ecological protection and energy accounted for an even lower share.

Even though amounts of Subject 211 for Environmental Protection increased continuously, the expenditure for environment is still less than other public services, such as rural development, education, medical and health, and scientific and technology development.

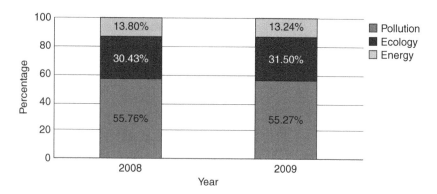

*Figure 2.10* Composition of Subject 211, 2008–2009.

Note
We calculated according to the data from the Ministry of Finance (2008–2009).

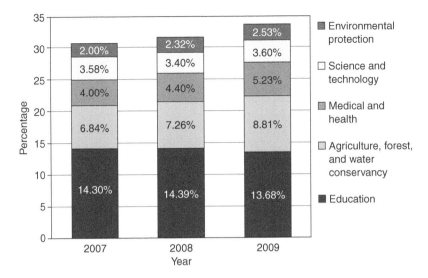

*Figure 2.11* Comparison of fiscal expenditures in China, 2007–2009 (source: Ministry of Finance 2007–2009).

The gap of fiscal expenditure for environmental protection remains large. For example, investments in environmental protection were RMB452.53 billion in 2009, only 42.7 percent of which were from fiscal funds. Therefore, it is rather urgent to increase the state's fiscal expenditure for environmental protection.

## Environmental taxation system of China

Among the 20 taxes of China's taxation system, there are several taxes associated with environmental protection, namely consumption tax, resources tax, value added tax, and income tax for enterprises.

### Character of the taxation system on environmental protection

Although the taxes described above have positive effects on environmental protection, no tax has been developed directly for the purpose of environmental protection. These taxes are the revenues prioritized rather than protecting the environment, such as consumption tax, urban maintenance and construction tax, vehicle and vessel tax. The resources related taxes are targeted at adjusting the differential incomes and promoting the rational utilization of resources, such as resources tax, land-use tax of cities and towns, and tax on occupation of cultivated land. The policies of tax preferences are to promote industrial restructuring and comprehensive utilization of resources, such as value added tax, and income tax.

No tax can focus on pollution control. The independent environment tax should be a tax on pollution behavior, which is imposed on the polluters who

Table 2.1 Incidence and stages of tax

| Tax | Incidence | Stages | | | |
|---|---|---|---|---|---|
| | | Exploitation and use | Production and sales | Consumption | Emission |
| Value added tax | Production and sales of products | — | ✓ | — | — |
| Consumption tax | Some products restricted by the state | — | ✓ | ✓ | — |
| Income tax | Production and management | — | ✓ | — | — |
| Resource tax | Mineral products and salt | ✓ | — | — | — |
| Urban and township land use tax | Urban and township land | ✓ | — | — | — |
| Farmland occupation tax | Cultivated land | ✓ | — | — | — |
| Vehicle and vessel tax | Vehicle and vessel | — | — | ✓ | — |

Source: Ma (2010).

have discharged the pollutants. The discharging behavior could be in the production or the consumption stage. As for the present environmental protection related taxes, the resources tax is levied in the mining stage, while urban land use tax and cultivated land occupation tax are levied in the resource using stage. Some items of the consumption tax are levied at the consumption stage. Yet there are no taxes that can be designed to be imposed in the discharging pollutants stage. As a result, an independent environmental tax is needed to charge at the discharging stage, so that all the stages from mining to using, production, consumption, and discharging will be completely under control.

Some taxes have the function to stimulate intensive utilization of resources as well as to reduce pollution. Some have raised funds for the prevention and control of pollution, which objectively helps protect the environment. The exemptions of some taxes such as value added tax, income tax, and export and import tax have positively contributed to reduce pollution, encourage investments in pollution control, decrease the operating cost of pollution control, and promote the development of the environmental protection industry. However, the lack of an independent environmental tax that could prevent actions of polluting and destroying the environment restricts not only the regulating efficiency against environmentally damaging behaviors, but also the sources of tax for environmental protection, which has weakened the effects of the related taxes.

It is impossible to introduce environmental tax in China by perfecting the present taxation system, or it would bring in huge changes for the taxation system. This is the reason why structural defects exist in the design of the current tax system. Therefore, establishing the new independent environmental tax and defining its prior and core position in the environment tax system would benefit the development of the entire tax system and will significantly help further perfect and deepen the reforms of the tax system.

### *The new environmental tax under discussion*

In the eleventh Five-Year Plan, the government set the targets for total emission control and energy saving. To achieve the goals, the State Council requested the use of legal, economic, technical, and administrative instruments to solve environmental problems. In the reform of resource tax, consumption tax, and import and export tax, environmental requirements should be fully taken into account. The State Council also requested to establish the environmental tax system, and use the tax leverage to promote a resource-saving and environment-friendly society. In June 2007 the State Council requested clearly "to study the introduction of environmental taxes". In 2010, the Central Committee of the Party proposed to continue the reform of fee to tax, completely reform the resources tax, and impose environmental tax in the twelfth Five Year Plan.

At the initial stage the environmental tax may target the emission of three pollutants: sulfur dioxide ($SO_2$), nitrogen oxides ($NO_X$), and wastewater. China's pollution levy system has been established for more than 30 years, which can provide technical support to improve performance of environmental taxes. China's environmental monitoring system also can play key roles in providing technical assistance to the environmental tax.

## China's ecological compensation

### *Major policies for ecological compensation*

More than 100 policies and regulations have involved ecological compensation up to now. In 1996, the State Council issued *Decisions on Several Environmental Problems,* which clarified that the compensation system for natural resource and ecological environmental restoration should be established and completed. The first law involving ecological compensation is the *Water Pollution Prevention and Control Law.* The first and the only special department rule on eco-compensation is the *Guidance to Carry out Ecological Compensation Pilot Work* by the former State Environmental Protection Administration (SEPA) in 2007. In 2009, the Ministry of Finance issued the notice on "transfer payments (the pilot) approach for national key ecological function areas", indicating that the transfer payments of the national key ecological function areas were established.

### *Ecological compensation practices*

According to the "guidance to carry out pilot on ecological compensation" by SEPA in 2007, the pilots on ecological compensation would be developed in four areas: nature reserves; important ecological function areas; mineral resources development aspect; basin water environment protection.

### *Ecological compensation for nature reserves*

By the end of 2009, 2541 nature reserves had been established in China (not including Hong Kong, Macao, and Taiwan), in which the number of national nature reserves is 319, accounting for 12.6 percent of the total. The reserve area covers about 147.75 million hectares, accounting for 14.7 percent of the country's territory. The area of the national nature reserves is 92.67 million hectares, accounting for 62.7 percent of the total area of the natural reserves.

A total of 73 percent of the nature reserve area is located in the western part of China where the economy is underdeveloped and the government's financial ability is limited. The nature reserve areas in Tibet and Qinghai cover 34.5 percent and 33.5 percent of the provinces' area respectively. In the eastern provinces the nature reserve area is less than 5 percent of the territory.

In 2009, the fiscal expenditure for the nature reserves was RMB5.368 billion, in which the central government covers 60 million, the financial transfer from the central to the local covers 1.327 billion, and the local financial expenditure covers 3.981 billion.

In 2001, the central government established a subsidy fund for forest ecological benefits, and the subsidy pilot areas covered more than 658 county-level units in 11 provinces and autonomous regions, and 24 National Nature Reserves, with average subsidies of RMB75 per ha per year, which provided a model for various types of nature reserves to establish ecological compensation mechanisms.

*An example: Sanjiangyuan Nature Reserve*

Sanjiangyuan Nature Reserve covers an area of 152,300 km$^2$, accounting for 21 percent of the total area of Qinghai Province, and 42 percent of the source area of the Yangtze, Yellow, and Lancang rivers.

According to the definition of the function zones specified by the eleventh Five Year Plan, Sanjiangyuan belongs to the development restricted and prohibited zone. In January 2003, the State Council approved the Sanjiangyuan Nature Reserve to be elevated to the national level. In 2008, the State Council issued the *Suggestion on Supporting Economic and Social Development in Qinghai and Tibet*, which explicitly required to speed up the establishment of the ecological compensation mechanism and to establish Sanjiangyuan comprehensive experimental zone of ecological protection in Qinghai. In 2009, the Ministry of Finance issued the *Notice on Transfer Payments Fund for the Ecological Function Zones (Sanjiangyuan, etc.)*.

The ecological compensation funds for Sanjiangyuan mainly come from two sources: the special ecological conservation fund planned by the *Overall Planning of Sanjiangyuan Ecological Protection and Construction*, and the general financial transfer payments.

According to the *Overall Planning of Sanjiangyuan Ecological Protection and Construction*, the total investment of ecological protection is RMB7.507 billion, in which the central government pays for RMB6.576 billion, and local government and herdsmen pay RMB0.931 billion. The annual investment from 2005 to 2010 was RMB1.5 billion per year, equal to RMB10,000 per km$^2$ and year.

## Comparative analysis of fiscal expenditure on education and environmental protection investment

The fiscal expenditure on education is the fiscal funds invested by the governments to support development of the educational system. The environmental protection investment refers to the funds for pollution control, ecological environment protection, and improvement (State Environmental Protection Administration 1999). It consists of national fiscal expenditure used for environmental protection and environmental protection funds paid by companies, individuals, and social groups.

### Policies relevant to fiscal expenditure on education

China's current policies relative to the fiscal expenditure on education mainly include three categories: laws, administrative regulations, and departmental rules. To ensure the scale of fiscal expenditure on education, the laws, policies, and plans, China set the target that the fiscal educational expenditure should account for 4 percent of GDP.

In 1993, the Central Committee of the China Communist Party and the State Council promulgated the *National Outline of Educational Development and Reform*, proposing that the fiscal educational expenditure should account for 4 percent of GDP by the end of 2000.

In 1995, the *Educational Law of People's Republic of China* announced that the proportion of fiscal educational expenditure in GDP should be increased gradually and the specific proportion could be prescribed by the State Council. In addition, the fiscal expenditures on education should be increased at a faster rate than the regular revenues.

However, in 2000 the proportion of fiscal educational expenditure in GDP was only 2.87 percent, even lower than that in 1986 and 1990. In 2001, the time to achieve the target was postponed to 2005. The *Compulsory Education Law of the People's Republic of China*, which was amended in 2006, stipulated that educational expenditure should be financed by government budget, and governments at various levels should make sure that expenditures on education increased at a faster rate than their regular revenues. In 2010, the *National Long-term Educational Reform and Development Plan (2010–2020)* postponed the target year to 2012. In 2012, China's GDP was RMB51.9 trillion and fiscal educational expenditure was RMB2.1 trillion, meaning that the proportion of fiscal educational expenditure in GDP was over 4 percent.

### Policies relevant to the environmental protection investment

The Regulation on Collection, Usage, and Management of Effluent Fees promulgated by the State Council in 2002 and Specific Measures for collection and use of the Funds from Effluent Fees put into effect in 2003, which emphasized that effluent fees must be included in the fiscal budget as special funds of environmental protection, among which 10 percent is derived from central budget revenue as central environmental protection specific funds and the remaining 90 percent stemming from the local budget revenues is contributed to local special funds of environmental protection.

In 2007, the SEPA and the National Development and Reform Commission (NDRC) issued the *Eleventh Five-Year Plan of National Environmental Protection*, proposing that the environmental protection investment accounts for about 1.35 percent of GDP.

In 2008, the proportion of environmental protection investment accounted for 1.49 percent of GDP, the only year to achieve the target during the period of 2006–2009. This was partly due to the global economic crisis in 2008 that slowed down China's economic growth. This proportion dropped to 1.33 percent in 2009.

Some legislative gaps exist in the field of environmental protection investment and the system of regulations and standards is not complete in China.

In 2008, the central government put forward the "4 trillion stimulus package" to maintain growth, including RMB210 billion in the field of energy saving, pollution reduction, and ecological construction projects.

## A comparative analysis

### Fiscal expenditure on education

In 2009, fiscal expenditure on education was RMB1.22 trillion, increased by 14.6 times with an average annual growth rate of 18.2 percent, compared with that in 1993, which was RMB86.776 billion.

Fiscal expenditure on education has been increasing since 2005, and the proportion in 2009 reached 3.59 percent, very close to the target of 4 percent.

The target of 4 percent has been set in accordance with the internationally accepted proportion of public expenditure on education in GDP as a measuring indicator of a country's level of government expenditure on education, as well as the proportion in developed countries (over 5 percent) and developing countries (4 percent). However, the time to achieve this goal has been delayed by at least 12 years in China.

### Amount of environmental protection investment

In 2009, environmental protection investment was 452.53 billion yuan, increased by 15.8 times compared with that in 1993, which was only 26.883 billion yuan, and its average annual growth rate was 19.8 percent, higher than that of fiscal expenditure (19.3 percent).

The proportion of environmental protection investment in GDP has been increasing and reached 1.49 percent in 2008, the only year to achieve the target of 1.35 percent, proposed in the eleventh Five-Year Plan of National Environmental Protection. However, it dropped to 1.33 percent in 2009 (Table 2.2).

According to the experience of developed countries, in the period of rapid economic growth, the proportion of environmental investment in GDP must maintain a steady 1–1.5 percent for a certain period (National Bureau of Statistics of China 2010), in order to effectively control the pollution. Only if it reaches 3 percent, the environmental quality will improved significantly. In view of this, China's environmental protection investment is insufficient.

It is estimated that during the twelfth Five-Year Plan demand for environmental protection investment in China will reach up to three trillion yuan, twice as much as that in the period of the eleventh Five-Year Plan.[1] Annual incremental expenditure will be 200 billion yuan so as to meet the demand for environmental protection investment, which is three trillion yuan.

China's fiscal expenditure was RMB10.4 trillion in 2011, but environmental protection is not included in the five major expenditures, which shows that the investment gap was rather great in environmental protection.

*Table 2.2* Total amounts and growth rate of fiscal expenditure on education and environmental protection investment, 1993–2009

| Year | Fiscal expenditure on education | | | Environmental protection investment | | |
|---|---|---|---|---|---|---|
| | Amount (100m) | In GDP (%) | Growth rate (%) | Amount (100m) | In GDP (%) | Growth rate (%) |
| 1993 | 867.76 | 2.46 | – | 268.83 | 0.76 | – |
| 1994 | 1174.74 | 2.44 | 35.38 | 307.2 | 0.64 | 14.27 |
| 1995 | 1411.52 | 2.32 | 20.16 | 354.86 | 0.58 | 15.51 |
| 1996 | 1671.70 | 2.35 | 18.43 | 408.21 | 0.57 | 15.03 |
| 1997 | 1862.54 | 2.36 | 11.42 | 502.49 | 0.64 | 23.10 |
| 1998 | 2032.45 | 2.41 | 9.12 | 721.8 | 0.86 | 43.64 |
| 1999 | 2287.18 | 2.55 | 12.53 | 823.2 | 0.92 | 14.05 |
| 2000 | 2562.61 | 2.58 | 12.04 | 1060.7 | 1.07 | 28.85 |
| 2001 | 3057.01 | 2.79 | 19.29 | 1106.6 | 1.01 | 4.33 |
| 2002 | 3491.40 | 2.90 | 14.21 | 1367.2 | 1.14 | 23.55 |
| 2003 | 3850.62 | 2.84 | 10.29 | 1627.7 | 1.20 | 19.05 |
| 2004 | 4465.86 | 2.79 | 15.98 | 1909.8 | 1.19 | 17.33 |
| 2005 | 5161.08 | 2.82 | 15.57 | 2388.0 | 1.30 | 25.04 |
| 2006 | 6348.36 | 3.00 | 23.00 | 2566.0 | 1.21 | 7.45 |
| 2007 | 8280.21 | 3.22 | 30.43 | 3387.3 | 1.32 | 32.01 |
| 2008 | 10,449.63 | 3.48 | 26.20 | 4490.3 | 1.49 | 32.56 |
| 2009 | 12,231.09 | 3.59 | 17.05 | 4525.3 | 1.33 | 0.78 |

Source: National Bureau of Statistics of China (2010).

Note
Data in the table are all present price.

Fiscal expenditure on environmental protection is rather small, compared with fiscal expenditure on education. For example, fiscal expenditure on education was 8.3, 7.2, and 6.3 times as much as that on environmental protection in the period from 2007–2009 (Figure 2.12).

Although the total amount of fiscal expenditure on environmental protection has been increasing year by year, its share in GDP and fiscal expenditure is still low. As a result, governments at all levels must continue to increase the total amount of fiscal expenditure on environmental protection.

### Constitution of fiscal expenditure on education

In the fiscal expenditure on education, budgetary educational funds take up the largest share. It reached about 1.2 trillion yuan in 2009, accounting for 97.9 percent and the proportion has been maintaining above 90 percent since 2004. Besides, the proportion of budgetary educational funds in fiscal expenditure has been increasing, average of which was 14.5 percent from 1993 to 2009 (Figure 2.13).

### Constitution of environmental protection investment

Environmental protection investment mainly consists of fiscal funds (such as budgetary funds and special funds), corporate self-financing funds, and bank loans. Corporate self-financing funds take up the largest share (more than 60 percent).

In 2007, the Subject 211 for Environmental Protection was first established as a composition of state fiscal budget. From 2007 to 2009, the share of fiscal funds

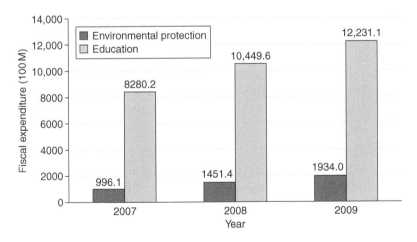

*Figure 2.12* Fiscal expenditure on education and environmental protection, 2007–2009 (source: National Bureau of Statistics of China 2010).

Note
Data in the figure are all present price.

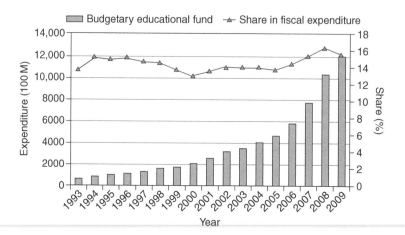

*Figure 2.13* Budgetary educational funds and its share, 1993–2009 (source: National
   Bureau of Statistics of China 2010).

Note
The budgetary funds contain the additional funding for education. Data in the figure are all present price.

in environmental protection investment was 29.4 percent, 32.3 percent, and 42.7
percent, respectively.

Owing to the small share of fiscal funds in environmental protection investment, environmental protection investment tightly depends on corporate self-
financing, and great investment gaps exist.

Among the total industrial pollution control investment in 2008, 95.8 percent
came from corporate self-financing, 1.6 percent was from effluent fees subsidies,
while other government subsidies contributed the remaining 2.5 percent in 2008.

In contrast to the fiscal expenditure on education, which mainly stems from
budgetary funds, environmental protection investment relies on corporate self-
financing, and the share of fiscal funds is lower than 50 percent in the long term,
leading to great gaps in environmental protection investment.

## Conclusion

The total amount of environmental protection investment and its share in GDP is
quite small, especially for fiscal expenditure on environmental protection, both
of which mainly depend on corporate self-financing rather than fiscal funds. This
has led to the pace of environmental management and recovery falling far behind
that of environmental destruction, bringing China's environmental protection
into a situation where "the previous outstanding debts has not been paid back
yet, while a new debt has been added".

Although education and environmental protection are both public services,
fiscal expenditures differ dramatically, suggesting government views education

more important than environmental protection. The reason may be the features of environmental issues, which are characterized by hysteresis, weak correlation of private interests, and uncertainty, relative to education. Environmental information asymmetry has also led to a lack of understanding of society, which is the main reason for them being more concerned about education with definite private benefits.

## Note

1 Ministry of Environmental Protection: demand for environmental protection investment in China will reach up to 3 trillion yuan in the next five years. Retrieved from: http://news.sohu.com/20101124/n277889677.shtml (last accessed June 18, 2013).

## References

Ma, Zhong (2010) *Environmental Economics and Policy: Theory and Applications*, Environmental Science Press, Beijing.

Ministry of Finance (2007–2009) The final statement of the national fiscal expenditure. Retrieved from: http://yss.mof.gov.cn/zhengwuxinxi/caizhengshuju/ (last accessed January 4, 2013).

National Bureau of Statistics of China (2010) *China Statistical Yearbook 2010*, China Statistics Press, Beijing.

National Development and Reform Commission (NDRC) (2009) Composition of the 4-trillion RMB investments and the recent progress of the projects invested by central government, May 21, 2009. Retrieved from: www.sdpc.gov.cn/xwzx/xwtt/t20090521_280383.htm (last accessed January 4, 2013).

National Environmental Protection Administration (1999) Establishment of environmental protection investment statistics system notice.

State Environmental Protection Administration (1999) Notice on establishing statistics system of environmental protection investment (Huan Cai Fa [1999] No. 64). Retrieved from: www.hb12369.net/download/115zhengzhikaohe.pdf (last accessed July 9, 2013).

Xie, Xuren (2008) *30 years of China's Fiscal Reform*, China Fiscal and Economics Publishing House, Beijing.

# 3 Re-visiting environmentally related taxes in China

*Jian Wu, Yujiao Mao and Zhong Ma*

In China, using tax instruments to facilitate energy conservation and pollution control has been widely discussed among academics for more than 15 years since the mid 1990s. Yet, only in the last five years has an environmental tax been seriously considered as a policy measure by the Chinese government. China's twelfth Five Year Plan (FYP) (2011–2015) clearly states the tasks of improving the resource and environmental tax system towards greening the finance and taxation system of the country. And now taxation instruments have been expected to play an important role in facilitating achieving the National Goal of Emission Reduction and Energy Conservation of China in the near future.

Against this backdrop, this chapter aims to evaluate the existing environmental tax system in China regarding its scale and composition. Drawing on our findings, we hope to bring some insights for policy-makers in designing and reforming environmental tax in China.

## Traditional understanding about environmentally related taxation in China

As many readers may know, there is no commonly agreed definition about environmental tax. In China, traditionally, the academic community tends to agree that all taxes with important environmental implications make up the overall environmental taxation system.

As is shown in Table 3.1, six environmentally related taxes have been included to form China's environmental tax system: natural resource tax, consumption tax (related items), urban construction and maintenance tax, vehicle and vessel tax, farmland occupation tax and land use tax. Consumption taxes are currently imposed on five energy-related products, including gasoline, diesel oil, motor tyres, motorcycles and cars. They are considered 'environmentally related' as they provide tax differentiation based on environment performance of these products. Property taxes applied to vehicle/vessel and land use are designed in a similar way. For example, the vehicle and vessel tax (enforced since January 1, 2012) set differentiated rates based on tonnage of ships and trucks, and type of cars. Urban construction and maintenance tax is

*Table 3.1* China's current system of environmentally related tax

| Category | Tax |
| --- | --- |
| Value added tax | VAT |
| Property tax | Vehicle and vessel usage tax (VVUT) |
| | Urban and township land use tax (UTLUT) |
| Turnover tax | Consumption tax (CT) |
| Resource tax | Resource tax (RT) |
| Income tax | Enterprise income tax (EIT) |
| Special purpose tax | City maintenance and construction tax (CMCT) |
| | Farmland occupation tax (FOT) |

Source: based on studies by Wang (2006), Wu and Xuan (2002), Jin and Zhou (2010) and Ge *et al.* (2006).

a surtax on the three major taxes (value added tax, consumption tax and business tax), and has become a major funding source for the construction and operation of urban environmental facilities (Jia and Wang, 2000). Value added tax (VAT) and enterprise income tax (EIT) have been characterized by many tax exemptions for environmental-friendly behaviors, so that the tax exemptions of VAT and EIT for environmental good are also considered as part of the environmental tax system, which, however, is not reflected in any calculation of the overall scale of the environmental taxes because the environmentally relevant part is difficult to be singled out.

Among these, the resource tax, which is now taxing on coal, petroleum, natural gas, other non-metal ore, ferrous metal, non-ferrous metal and salt, is expected to be the one with the most explicit environmental dimension. However, the main purpose of this tax is not to promote conservation or sustainable use of natural resources, rather it is designed to adjust the income of companies and promote market competition. Moreover, the current resource taxes only focus on non-renewable resources and do not cover renewable resources (forests, water, etc.).

With this definition, according to the statistics during 2002 to 2009, revenues from environmentally related taxes are small, but not insignificant, averaging around 0.84 percent of gross domestic product (GDP) and 5.05 percent of total tax revenue, as given in Figure 3.1.

During 2002 to 2009, the revenues from environmentally related taxes kept a continuous and steady rising tendency, and recorded a sharp growth rate in 2009, as shown in Figure 3.1. This is due to the increase of the oil consumption tax in 2009. Figure 3.1 further displays that the ratio of the revenues from environmentally related taxes to GDP and total tax revenue showed a continuous, steady upward trend from 2002. Up to 2009, the environment-related tax has accounted to 1.8 percent of GDP, and 10.7 percent of overall tax revenue.

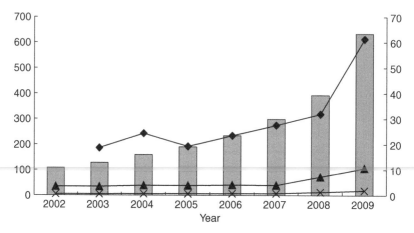

*Figure 3.1* Revenues from environmentally related taxes, 2002–2009[1] (sources: National Bureau of Statistics 2003–2010; National Bureau of Taxation 2003–2010).

Note
1 The index's statistic specifications are as follows: excise Goods and Services Tax (CT) (deducting the irrelevant revenues of tobacco and alcohol and treating the remaining part as income related to environmental protection), resource tax, vehicles and vessel tax, pollution levy, urban and township land use tax, and tax on the occupancy of cultivated land; vehicles and vessel tax before 2007 is the sum of vehicles and vessel use tax and vehicles and vessel license use tax.

## Re-visiting environmentally related taxation in China

### Environmentally related tax bases

The environmentally related taxes can also be defined and calculated in another way. The OECD, the International Energy Agency (IEA) and the European Commission have agreed to define environmentally related taxes as any compulsory, unrequited payment to general government levied on tax bases deemed to be of particular environmental relevance (OECD, 2010). The relevant tax bases include, for instance, energy products, motor vehicles, waste, measured or estimated emissions, and natural resources.

Similarly, in China environmentally related taxes based on the above tax bases can be classified into seven categories: tax on transportation fuel, tax on heating and processing fuels, electricity tax, tax on motor vehicle, natural resource tax/fee, other environmentally related product tax and pollution levy.

This definition constitutes the environmental taxation system by screening the tax bases that are relevant to environmental effects, based on which to decide whether a tax should be included as an environmental tax or not, and then to evaluate the scale and composition of environmental tax. This is more essential for the definition of environmental tax. As we know, taxes may have been implemented for a number of reasons, most likely general revenue-raising, with little to no consideration for the environment. Moreover, some taxes have been implemented without carefully assessing the costs and damages of the pollution, leading to non-optimal rates. Therefore, it is appropriate to consider only the type of tax base and to include the revenue from certain tax bases as complete as possible, not the intention or appropriateness of the instrument.

Moreover, by doing so, all taxes levied on a certain tax base will be counted, including VAT, because its tax bases include those which may be related to the environment. Regarding on the debate on whether VAT on environmentally related production or products should be counted in as environmental tax, some studies would support to count it in so that some environmental tax base categories such as electricity and heating/processing fuels can be counted in (Man *et al.*, 2010), while other studies would support to exclude VAT from the definition of environmental taxes, because it does not influence relative prices in the same way that other taxes on environmentally related tax bases do since that VAT is a tax levied on all products (with few exceptions), and deductible for many producers (European Communities, 2001).

Although we notice that there are cases in other countries that VAT rates vary according to environmental factors, for example Austria and Spain used to have special high VAT rates on vehicles (European Communities, 2001), China's VAT does not charge different rate on different items, not like consumption tax. China does have VAT deduction or exemption policy to support environmental-friendly production activities, which makes impacts on the relative price of the products, but it is difficult to separate their scale from the overall statistics.

In this chapter, we tend to count on the VAT on environmentally related items (tax basis) to fully reflect the tax burden on environmentally related bases in China. This VAT impacts the final price of most products, although not the relative price of products, but is still influencing consumer's behavior if only the price elasticity still make effects. The second reason is that China's taxation structure has been designed to levy higher taxes on production and turnover processes, rather than on consumption or income. By including the VAT of a product, we are able to count all taxes on environmentally related tax bases in the entire production, sales and consumption chain.

In this chapter, in order to avoid any overestimation in comparing with other countries, we drop all the VAT from the total environmentally related revenue when we compare China with other countries.

### *Environmentally related tax in China: issues of scale*

Since urban maintenance and construction tax is traditionally taken as an environmentally related tax, we will also include it in the following calculation. Therefore, in this new scope, there are eight categories of environmentally related tax in China (Table 3.2).

*Table 3.2* Environmentally related taxes under new scope

| Tax bases | Taxes |
| --- | --- |
| Transportation fuels | VAT on refined oil product |
| | CT on refined oil product |
| Heating and processing fuels | VAT on coal, coke |
| | VAT on crude oil, natural gas |
| | VAT on gas |
| Electricity | VAT on electricity production |
| | VAT on electricity transmission |
| Motor vehicles | Vehicle excise duty |
| | Vehicles and vessel tax |
| | VAT on automotive |
| | VAT on motorcycle |
| | CT on automotive |
| | CT on motorcycle |
| Natural resources | Resource tax |
| | Urban and township land use tax |
| | Farmland occupation tax |
| | VAT on mineral |
| | VAT on water |
| | Water resource fee |
| Other environmentally related product tax | VAT on tires, 17% |
| | VAT on manure, 13% |
| | VAT on pesticide, 13% |
| | CT on tires, disposable chopsticks and wooden floor |
| Pollution | Pollution levy |
| Surtax | Urban maintenance and construction tax |

Source: National Bureau of Taxation (2010).

Derived from statistical data from 2007 to 2009, the revenues from environmentally related tax show a continuous, steady rising tendency, with the growth rate of revenues also increasing rapidly (Figure 3.2). Revenues from environmentally related tax reach RMB792.46 billion, RMB988.13 billion and RMB1,253.67 billion, in 2007, 2008 and 2009, respectively. The year in 2009 stands out with a significant increase in its revenues, which is relevant to the comprehensive rise on consumption tax rate of refined oil product.

Generally, during 2007 to 2009, the revenue from environmentally related taxes averages around 3.33 percent of gross domestic product (GDP), and accounts for 18.87 per cent of total tax revenue.

### Environmentally related tax in China: composition

As regarding the composition of eight categories of environmentally related taxes, tax revenue on transportation fuel increased from RMB59.71 billion in 2007 to RMB247.23 billion in 2009. Its share among total environmental tax revenue has increased from 7 to 20 percent during 2007–2009. The sharp rise of

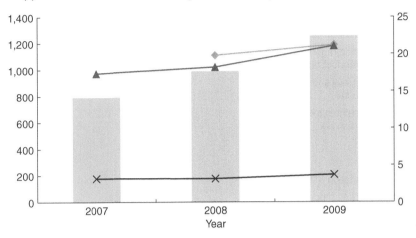

*Figure 3.2* Revenues and growth rate of environmentally related tax in China (sources: National Bureau of Statistics, 2008–2010; National Bureau of Taxation, 2008–2010; Ministry of Environmental Protection, 2008–2010a, 2008–2010b).

the consumption tax rate in 2009 can explain why the revenue in 2009 is five times larger than that of the previous two years. The adjustment is an important step of China's refined oil price and tax reform, which is of great importance to promote energy conservation and emission reduction, and to reform the fee system in China. The composition of revenues rose from eight environmentally related taxes during the period of 2007–2009 (Figure 3.3).

China doesn't have an electricity tax, but the government imposes a tax on electricity production and electricity transmission through the VAT. The revenue from the electricity tax was 155.97 billion yuan in 2007, accounting for 20 percent of revenue from environmentally related tax, and slightly decreased to 147.47 billion yuan, accounting for 12 percent of revenue from environmentally related tax.

The revenue from the tax on heating and processing fuels is 141.06 billion yuan in 2007, covering 18 percent of total environmental revenue, and increases to 181.41 billion yuan in 2009, 14 percent of total environmental revenue.

The tax on motor vehicles includes one-off taxes (vehicle excise duty, VAT and CT), and recurrent tax (vehicles and vessel tax). The revenue from the tax on motor vehicles reached 169.14 billion yuan, accounting for 20 percent of the environmental tax in 2007, and kept increasing to 262.66 billion yuan in 2009, accounting for 21 percent of environmental tax, mainly due to the sharp increase in motor vehicle consumption. This revenue is expected to further increase in 2012 after the new Law on Vehicle and Vessels has come into force.

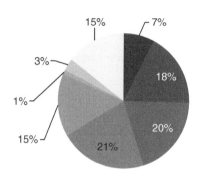

- Transportation fuels
- Heating and process fuels
- Electricity
- Motor vehicles
- Natural resources
- Other environmentally-related product tax
- Pollution levy and water resource fee
- Urban maintenance and construction tax

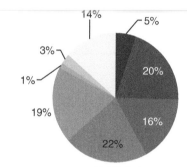

- Transportation fuels
- Heating and process fuels
- Electricity
- Motor vehicles
- Natural resources
- Other environmentally-related product tax
- Pollution levy and water resource fee
- Urban maintenance and construction tax

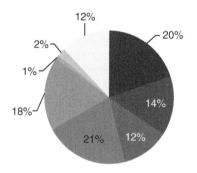

- Transportation fuels
- Heating and process fuels
- Electricity
- Motor vehicles
- Natural resources
- Other environmentally-related product tax
- Pollution levy and water resource fee
- Urban maintenance and construction tax

*Figure 3.3* Revenues raised on environmentally related tax in China, 2007–2009 (sources: National Bureau of Statistics, 2008–2010; National Bureau of Taxation, 2008–2010; Ministry of Environmental Protection, 2008–2010a, 2008–2010b).

Revenues from natural resources are 125.74 billion yuan in 2007, accounting for 15.87 percent of overall environmentally related revenue, and 231.81 billion yuan in 2009, accounting for 18.49 percent. This tax revenue is also expected to be greatly increased since the new Natural Resource Tax reform on November 1, 2011, by which the tax base for crude oil and natural gas has been modified from previously levying on volume to levying on rate (5 percent). The increase is predicted to be 12.6 times the figure before the reform (Jin and Wang, 2010).

The pollution levy/fee was 17.36 billion yuan, with a ratio of 2.19 percent to environmentally related tax in 2007, and 17.26 billion yuan with a ratio of 1.38 percent in 2009. The pollution levy is a quasi-tax based on emission or discharge and includes sewage charges, air emission charges, solid waste and hazardous waste charges, and excessive noise charges. In recent years, the pollution levy has seen a decreasing tendency due to the pollution reduction from monitored and regulated pollution sources, as is the case for air emissions, or because more and more sewage water has been diverted to wastewater treatment plant, on which the pollution fee is no longer applied.

Urban maintenance and construction tax was 115.64 billion yuan, 15 percent of the environmental tax revenue in 2007, and has been increased to 154.41 billion yuan in 2009, accounting for 12 percent of environmental tax revenue. In addition, shares of other environmentally related product tax are small, always less than 1 percent.

In general, up to 2009, tax on motor vehicle and transportation fuels has accounted for the largest share of almost 41 percent in the overall revenues from environmentally related taxes, followed by the taxes on natural resources, heating and processing fuels, urban maintenance and the construction tax. The pollution levy is very small.

Figure 3.4 shows the ratios of specific environmental tax to GDP. Similarly, the largest proportion is the tax on motor vehicles, which increases annually; tax on transportation fuels decreases first and then increases in 2009, whose share accounts for the second largest; there is a similar proportion of the tax on heating and processing fuels, electricity tax and natural resource tax. The ratio of natural resource tax increases annually; the proportion of electricity duty and heating and processing fuel tax gradually declines, which can be explained and predicted by economic fluctuation trends. In addition, the waste tax, other environmentally related product taxes and the pollution fee have been maintained at a relatively low but generally stable share. As seen in Figure 3.4, the vast majority of environmentally related tax revenues are derived from taxes on energy, natural resources, and vehicles and vessels.

## Discussion

The fundamental difference between the traditional understanding and the revisiting one is whether to count VAT in. Besides, the second understanding is much clearer to reveal the source of revenue.

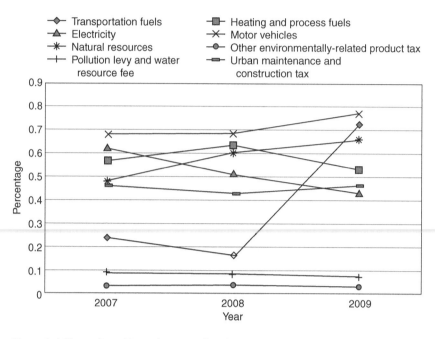

*Figure 3.4* Share of specific environmentally related taxes in GDP, 2007–2009 (%) (sources: National Bureau of Statistics, 2008–2010; National Bureau of Taxation, 2008–2010; Ministry of Environmental Protection, 2008–2010a, 2008–2010b).

If we would understand the full burden of environmentally related tax, VAT has to be counted in. As is shown in Figure 3.2, environmentally related taxes accounted for a quite large portion of 18.89 percent within the government's revenue, and 3.33 percent of GDP. In terms of the composition, the tax on motor vehicles is the largest environmentally related tax revenue source.

If compared with other OECD and European countries where VAT has been excluded from statistics of environmental taxes, we can drop the VAT from the above calculation. The revenue of environmentally related taxes during 2007–2009 is 1.79 percent of GDP, and 10.13 percent of total fiscal revenue, which can be compared with the OECD average level of environmentally related taxation in 2008, averaging around 1.5–2 percent of GDP, and 5–6 percent of total revenues for governments.[1] And, of course, environmentally related taxes in China have contributed to a larger portion to the government revenue, compared with OECD countries. Excluding VAT will reduce the scale of environmentally related tax by a half, and some categories will disappear from the package, such as tax on heating and processing fuel, and electricity (Figure 3.5).

As for the composition, taxes on transportation energy have become the largest source of environmentally related tax revenues, similar to the OECD countries, where a vast majority of the environmentally related tax revenues come from energy taxes, accounting for about two-thirds of the total environmental revenues, and tax on motor vehicle and other transportation covering one-third. But the share of tax revenue from transportation energy and motor vehicles in China is still much smaller (about 40 percent) than OECD countries (over 90 percent).

## Conclusions

Taking a fresh look at the environmentally related tax in China, we find the following surprising facts.

First, environmentally related taxes have already been developed into a certain scale that is comparable to other countries. If VAT is counted in, these taxes have already become very meaningful tax revenue sources for the government, and, at the same time, accounted for a very important proportion to the overall tax burden.

Second, China's structure of environmentally related taxes relies increasingly on the major tax bases such as energy products, as in the OECD countries. This implies that these taxes will continue to be significant in terms of revenue raising function, as other countries' experiences have proven that energy products are the only tax base that will be meaningful for revenue-raising among all other environmentally related tax bases, since they will not necessarily be regressive.

So far China has set up a system of environmentally related taxes, and there is room to increase further the scale of these taxes by future reforms, especially on natural resources tax, energy tax and tax on motor vehicles. It is the right time for China to think about how to use these revenues efficiently to achieve higher

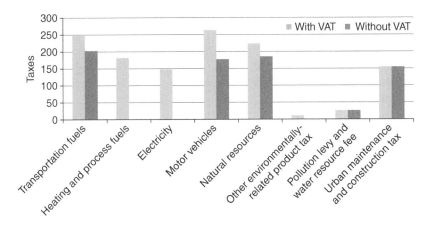

*Figure 3.5* Environmentally related taxes with and without VAT, 2009 (sources: National Bureau of Statistics, 2008–2010; National Bureau of Taxation, 2008–2010; Ministry of Environmental Protection, 2008–2010a, 2008–2010b).

social welfare. Greening the overall tax system under the revenue neutral principle, as is shown by the experience of the OECD countries, should be considered for China's future tax reforms.

## Acknowledgments

This study was supported by China's Key Project of National Social Science Funding: China's Environment Taxation System (No. 09&ZD052). We owe our gratitude to Dr Stefan Speck from the European Environment Agency for his generous support during our data collection, and we appreciate the valuable comments from reviewers of this chapter.

## Note

1 In OECD's calculation, broad-based taxes, such as VAT are not included as environmentally related taxation, mainly because VAT differs substantially between countries.

## References

European Communities (2001) *Environmental Taxes: A Statistical Guide*, Luxembourg: Office for Official Publications of the European Communities.
Ge, C.Z., Wang, J.N. and Gao, S.T. (2006) *Environmental Taxation and Public Finance*, Beijing: Environmental Science Press of China.
Jia, K. and Wang, G.J. (2000) 'Discussions about improving environmental taxation in China', *Taxation Research*, No. 9: 43–8.
Jin, D.S. and Zhou, H.W. (2010) 'On the features, problem and reform of China's resource tax system', *Taxation Research*, No. 7: 40–4.
Man, Y.Y., Zheng, X.Y. and Zheng, Y.E. (2010) 'China's environment tax: from an international visual angle', *Peking University – Lincoln Institute Research Briefings*, No. 1: 34–40.
Ministry of Environmental Protection (2008–2010a) *China Statistics Annul on Environment*, Beijing: Environmental Science Press of China.
Ministry of Environmental Protection (2008–2010b) *China Statistics Bulletin on Environment*, Beijing: Environmental Science Press of China.
National Bureau of Statistics (2003–2010) *China Statistic Yearbook*, Beijing: China Statistics Press.
National Bureau of Taxation (2003–2010) *China Taxation Yearbook*, Beijing: China Tax Press.
OECD (2010) *Taxes, Innovation and the Environment*. Online, available at: www.oecd.org/publishing/corrigenda (accessed January 5, 2012).
Wang, J.N. (2006) *Environmental Taxation and Implementation Strategy*, Beijing: Environmental Science Press of China.
Wu, Yajun and Xuan, Xiaowei (2002) *Theory of Environmental Tax and Practice in China*, Beijing: Economic Science Press.
Wu, Y.J. (2005) 'Some theory and empirical problems about greening Chinese taxation', *Economical Science*, No. 1: 77–90.

# 4 Political economy of implementing pollution control fees

## The case of Taiwan

*Li-Chun Chen*

## Introduction

In Taiwan, prior to the 1970s, pollution management referred to simple control of pollution through regulatory measures. Examples include the specification of emissions standards, regulation of emissions, and creation of permission systems. For this reason, the social cost induced by environmental pollution was not reflected properly in policies, and an enormous amount of labor and other resources were wasted in policy execution. Also, the effect of these regulatory measures was limited because of a lack of incentives for polluters to proactively contribute to the reduction of pollution. Stemming from environmental awareness, most environmental acts are not only progressive in spirit, but also integrate modern environmental control mechanisms. However, the effect of these regulatory measures has been limited, since the economic incentive in which the polluters positively improve the pollution status is lacking. Therefore, the Taiwanese government has examined the environmental policy systems of European and American countries and Japan since the 1990s, and now is introducing the economic instrument under green budgeting approaches. This led to Taiwan's pollution control fee systems.

Taiwan's current tax system already contains some elements that are closely linked to protecting the environment. For example, the commodity tax, vehicle license tax, fuel fee, anti-pollution fees imposed by the Environmental Protection Administration, and fees for mineral exploration and mining, water conservation, and compensation. Revenue from these taxes and fees in 2008 exceeded NT$220 billion, representing over 1.75 percent of GDP. In comparison, environmental taxes among OECD countries range from 2 to 2.5 percent of GDP, so there is clearly room for greater effort. In addition to taxes and fees, Taiwan also has a number of environmental tax preferences, such as: 50 percent commodity tax reduction for electric vehicles; accelerated depreciation and investment tax credits for energy-saving or clean/alternative energy-using machinery and equipment; and reduced land value, estate, and gift taxes for land in designated watersheds. While green taxes, fees, and preferences involve different administrative turfs and budgets, as well as separate policies for environmental protection, industrial development, and revenue collection, Taiwan's government is now pressing ahead with a new round of tax reform, and the Statute for Upgrading Industry is set to expire at the end of the year, so this seems like a good time to offer some humble opinions on building a green tax system and encouraging industrial transformation.

Among Taiwan's pollution control fees, the fee for general waste disposal and the recycling fee have been studied recently by Murakami (2006) and Nanbu (2007). This chapter examines the political economy of the air pollution control fees (hereinafter referred to as the APC fees) which have not yet to be sufficiently analyzed.

This chapter is structured as follows. The next section discusses the context surrounding the introduction of pollution control fees in Taiwan and the structure of the fee system. Then the role of stakeholders in the implementation process is examined. Last, the resulting effects of APC fees and issues to be considered are outlined.

## Overview of Taiwan's pollution control fee systems

The idea that environmental taxes and levies are more cost efficient than environmental regulations in reducing pollutants is supported by many, including Pearce *et al.* (1989), von Weizsaeker (1990), Uzawa (1995), and the OECD (2001).

Taiwan first introduced an economic instrument into its environmental policy in early 1987. Sixty-four legislative members jointly signed the Draft of the Environmental Protection Basic Act, submitted it to parliament (Legislative Yuan), and began deliberation. Article 21 of the Draft stated that the central administrative agency should set the rate of pollution control fees based on the degree of pollution and collect them. In the same year, the administration (Executive Yuan) published the Current State of the Environmental Protection Policy Principles which stated in Chapter 3, Section 6, that a system needed to be established to make polluters bear costs, and that the government should collect pollution control fees from polluters in addition to making them use pollutant-reducing equipment and contribute to the improvement of the environment. Article 9 the Draft of the Environmental Protection Basic Act, which was a deliberation proposal by the Executive Yuan submitted in the following year to the Legislative Yuan, also stipulated the creation of a system which would impose costs on polluters. Today, the deliberation on the enactment of the Environmental Protection Basic Act has yet to be completed. However, the economic instruments indicated in both the Current State of the Environmental Protection Policy Principles and the Draft of the Environmental Protection Basic Act has been gradually realized in various laws addressing environmental protection.

Although Taiwan has not established a system of environmental taxes and pollution control fees that are defined by the OECD, there are equivalents— energy taxes, pollution control fees, natural resource taxes, and transportation taxes. All of the pollution control fees collected by the Taiwan Environmental

Protection Administration (TEPA) are revenues earmarked for limiting the source of pollution or properly processing relevant pollutants. The energy taxes administered by the Ministry of Finance, Ministry of Economic Affairs, and Ministry of Transportation and Communication are revenue sources earmarked for mainly the reduction of air pollution and road constructions. The natural resource taxes and transportation taxes are general revenue sources for which no specific uses are predetermined.

The share of the energy taxes, pollution control fees, natural resource taxes, and transportation taxes in the environmental taxes for 2006 was 45.6, 4.8, 0.1, and 49.5 percent, respectively. As to the earmarked revenue sources, namely, the energy taxes and pollution control fees, the share of the former in the total tax revenue for 2004, 2005, and 2006 was 10.8, 9.9, and 10.5 percent, whereas the share of the latter for the same years was small at 1.3, 1.2, and 1.1 percent, respectively. The sum of the two revenue sources (12.1, 11.1, and 11.6 percent) is much smaller than the income tax revenue (32.9, 39.9, and 40.4 percent, respectively) (Directorate General of Budget, 2007; Ministry of Finance, 2008).

As for the pollution control fees, the following are levied under a corresponding law: the APC fees are levied under the Air Pollution Control Act; the fee for general waste disposal and recycling fee under the Waste Disposal Act; water pollution control fees under the Water Pollution Control Act; the fee for soil and groundwater remediation under the Soil and Groundwater Remediation Act; the fee for man-made marine pollution (effluent or contamination) under the Marine Pollution Control Act; and the airplane noise control fee under the Noise Control Act. Each of these pollution control fees enters a fund as earmarked revenues, and the revenues are used for measures to protect the environment. TEPA recently created the Environmental Protection Fund which combines four major funds—the APC fund, recycle management fund, soil and groundwater remediation fund, and the water pollution control fund.

## Process of introducing the APC fees and the fee structure

The Air Pollution Control Act (APC Act), which was amended in 1992, prescribes the collection of fees associated with reducing air pollution. The process toward creating a system to control air pollution took a long time and involved policy planning and discussions, hearings with experts, the assessment of the effectiveness and benefits of implementing the system, discussions in academic circles, and public participation in discussions. The process faced intense questions and fierce protests by stakeholders opposing the new policy, but moved to the first phase of implementing APC fees on July 1, 1995 with the formal initiation of the sulfur oxide (SOx) APC fee. The collection of other fees followed: the building construction particulate APC fee started in July 1997; the launch of the nitrogen oxide (NOx) APC fee in July 1998 marked the second phase; and the launch of the volatile organic compounds (VOCs) APC fee on January 1, 2007 marked the third phase.

## Policy-making process

The Air Pollution Control Fee Levy Bill prepared in 1991 considered both non-point and point sources of pollution.[1] Non-point sources included automobiles and motorcycles. Point sources varied in their scope, industry type, and discharged pollutants, but the steel industry and cement industry were initially considered as research subjects. The level of pollution was estimated for 1993 and 1998 based on the amount of emissions and the number of pollution sources. The base charge rate as well as the selection and calculation of the rate played an important role in establishing the pollution control fee system.[2] As for the point sources, the base value was the level of pollution per unit of production duration (ton/month) for the steel and cement industries. The cost of controlling pollution was taken into account in calculating the charge rate, and also the rate was designed to be adjustable so that it could be changed at appropriate times.

The collection of the APC fees was officially stipulated with amendment of the APC Act in February 1992. Article 10, Paragraph 1 states that APC fees be collected at any administrative agency level based on the type and quantity of pollutants discharged into the atmosphere; Paragraph 2 states that central government agencies and relevant entities determine the type of pollutants and methods for fee collection mentioned in the previous paragraph. Also, Article 14 prescribes the implementation of this law in detail: Paragraph 1 states that the APC fees be collected at any administrative agency level according to the rules in Article 10 of the law, and that the collected resources are to be used specifically for reducing air pollution; Paragraph 2 stipulates that a fund can be established and administered with the collected resources mentioned in the previous paragraph, and that its revenue and expenditure as well as administrative methods would be determined separately by administrative agencies. Article 6 states that provincial, county, or city administrative agencies should prepare plans to maintain and improve the air quality of their jurisdiction based on air quality standards.

In order to make the levy system for the APC fees compatible with the efforts of each county or city to promote its plan to maintain and improve air quality, TEPA prepares plans to actively use the APC fees and tries to achieve its goals of improving air quality. On November 30, 1994, TEPA drew up a draft of the Proposal for the Implementation of Plans to Improve and Maintain Air Quality and the Use of Air Pollution Control Fees and submitted it to the Legislative Yuan for deliberation. Corrections were made by March 3, 1995, and the proposal was approved by the Executive Yuan on May 1 of the same year.

## Policy developments

The creation of the APC fee system started immediately following the amendment to the APC Act which was made in 1992. After numerous Ministry- and Commission-level discussions and meetings held in 1992 and 1993, there were opinions that the collection of APC fees would influence the political circumstances of the time. Hence, TEPA proposed in May 1993 that the Executive Yuan postpone the implementation of the fees, but the Executive Yuan ordered the devising of implementation methods to be continued. In response, TEPA held the first public hearing on January 12, 1994 and, on January 22 of the same year, invited representatives from relevant agencies to cooperatively design the collection method.

A public hearing held by TEPA and the draft proposal for the collection of APC fees caused widespread objection from various organizations and citizens. There was so much opposition at the time that TEPA decided to temporarily withdraw the draft proposal submitted to the Executive Yuan. TEPA held several public hearings and seminars in March and April of 1994, sorted out opinions gathered from different groups, and submitted a revised draft proposal for fee collection to the Executive Yuan in June. The draft was approved in March 1995 and was made public by TEPA on March 23. The 1996 budget proposal for the entire central government included the APC fund.

During the deliberation process, stakeholders from a variety of fields showed strong interest in the mechanics of the APC fee system. A wide range of issues were discussed during the initial stage of the process regarding whether the collected fees would be treated as general purpose, or special purpose, revenue. Many proposals on this question were submitted in February 1994 during the third session of the Second Legislative Yuan. Some legislative members supported the idea of the polluters-pay principle, but disagreed with its actual implementation.

During the deliberation on the budget associated with the APC fund, various opinions were expressed by organizations for consumer protection or environmental protection. There were also proposals to eliminate such budget; there was a call, jointly signed by Legislator S and others, to remove the draft proposal for the collection of APC fees aimed at non-point sources, and an opinion, jointly signed by Legislators S and K and others, to postpone the collection of APC fees.[3]

The APC fees create not only a legislative problem, but also a sensitive political issue. Due to intensified social movements and the process of democratization in Taiwan, opposition against the additional burden imposed by new public policy had strengthened, which made it difficult to implement a system of collecting fees. Furthermore, the implementation of the fees was also facing challenges from environmental protection organizations and was influenced by conflicts between the ruling party and others in the Legislative Yuan. Given the social and economic conditions of the time, it was considered that, although their economic influence would be small, the collection of the APC fees would have a substantial impact politically without appropriate measures being taken. Therefore, it was decided that TEPA proactively hold, in stages, policy briefing sessions for legislative members in order to secure financial resources for improving air quality efficiently based on the polluters-pay principle.[4]

On July 1, 1995 the first phase of the APC fee collection officially started. Groups that maintained strong opposition held protests on the morning of that day by hanging banners expressing their objection at ten gas stations in nine counties and cities, including one on Jian-guo South Road in Taipei, making public speeches, and requesting people to reject the fee collection. However, despite such protests, the rejection represented merely one-hundred-thousandth of the volume of daily transactions at 1,179 gas stations nationwide. Therefore, the system of the APC fees was officially implemented receiving support from people and various industries.

### Changes in the fee collection method

The following represents the main content of the initial draft proposal regarding the method of collecting the APC fees: (1) the polluters declare the amount of their pollutant emissions; (2) the charge rate is determined according to the type of pollutant; (3) the local governments have authority to set a charge rate higher than the TEPA benchmark (enabling proper responses to various conditions of different regions); (4) the charge rate can be adjusted to the total amount of pollutants (variable charge rate); and (5) the local administrative agencies collect the APC fees from both the point and non-point sources.

Article 10 of the APC Act amended in February 1992 stipulated that the APC fees be collected at any level of administrative agencies based on the type and quantity of pollutants discharged into the atmosphere. After several public hearings and numerous discussions and consultations with industries, a method for collecting fees was made public in March 1995. In July of the same year, the collection of the SOx APC fee from the point sources started under two categories: the fee based on the amount of fuel use and the fee based on sulfur content. Later, to comply with the interpretive document from the Council of Grand Justices, the collection method which was revised in May 1998, shifted to the collection of fees based on actual emissions of pollutants.[5] This included the collection of the NOx APC fee.

To spread the use of the polluters-pay principle, the second phase of the APC fees introduced the collection of both the SOx and NOx APC fees from the point sources starting in July 1998 based on the actual amount of pollutants discharged by them. This expanded the effect of the economic instrument and contributed to the reduction of pollutant emissions while making the point pollution sources take optimal and efficient measures to limit their emissions. Also, the third phase of the APC fees started on January 1, 2007 with the inclusion of the VOCs APC fee.

### Calculation of the amount of emissions

The declaration of the amount of emissions can be done using calculations based on numerical results of continuous automated monitoring, quantities obtained through sampling, or predicted values of emission coefficients. This calculation method is employed most among all the calculation methods (71 and 77 percent, respectively). This is followed by the method based on sampling (25 and 20 percent).

## Charge rate and discounts

According to the provision in Article 17 of the APC Act, the charge rate used for APC fees is set based on existing air quality conditions, sources of pollution, pollutants, types of oil (fuel), and costs of reducing pollution. The local administrative agency in charge of a district that is under emissions regulations evaluates the condition of the environment and air quality of the district regularly after one year of implementing the charge rate and can give the central administrative agency a proposal to adjust the charge rate within 30 percent above or below the rate set in the previous paragraph. When the central administrative agency reviews and approves the proposal, the local administrative agency announces the adjusted charge rate.

TEPA raised the charge rate for the VOCs APC fee and also began calculating emissions by firms and individuals to obtain the basis for setting the charge rate for both the SOx and NOx APC fees when it announced the Charge Rates for the APC Fees from Point Pollution Sources on December 27, 2006. This was done not only to provide polluters with incentives to reduce NOx emissions, but also to incorporate effectively the marginal cost of reducing pollution. The calculation of the charge rate was divided into three stages. Also, until December 31, 2009, the VOCs APC fee was obtained by multiplying the amount of emissions above one ton (which is the amount of deduction allowed for each quarter) by the charge rate (20 yuan/kg). But, similarly to the case for SOx and NOx, the VOCs APC fee is subject to additional economic instruments, namely, differentiated charge rates and discounts.

A preference coefficient is newly introduced to provide firms and individuals an incentive to reduce emissions of pollutants. It is applied as follows. A preferential treatment—a discount of 20 to 60 percent—is given to those who reduce SOx and NOx emissions by installing pollution control equipment or improving their production process and emit SOx and NOx with the level of concentration being below 100 ppm and to those who can reduce their NOx emissions to below 50 percent of the standard level. Moreover, an additional 10 percent discount is given to those who can reduce their pollutant emissions effectively, set up equipment that reduces emissions of harmful substances, or keep the concentration of emitted pollutants below 50 ppm. With these measures, firms and individuals can receive a discount of up to 64 percent. Also, in order to reduce NOx emissions effectively, a further preferential treatment—an additional 40 percent discount—is applied to qualified firms and individuals when their pollution sources are equipped with a selective catalytic reduction (SCR) system and emit pollutants whose concentration is below 40 ppm. If this is applied, firms and individuals can receive a discount of up to 76 percent for their NOx APC fee (Table 4.1).

## Fee collection from non-point sources

As for non-point sources, both the draft proposal for the fee collection method and the content of the 1995 fee implementation announcement set the amount of

Table 4.1 Calculation method for a gross preference coefficient and the conditions of applicability

| Rank ratio (A) | Preference coefficient (A') | Notes |
|---|---|---|
| A≥95% | 40% | ① Rank ratio (A) = emission under applicability conditions/total emission (B) |
| 75%≤A<95% | 50% | ② Rank ratio (A)≥30% and flue concentration of 50 ppm or below |
| 50%≤A<75% | 65% | 10% discount applied to emissions through such flues |
| 30%≤A<50% | 80% | Gross preference coefficient (D) = [(C1/B) × 90% + (C2/B)] × A' |
| | | A': preference coefficient |
| | | B: total emission |
| | | C1: emission under applicability conditions and concentration ≤50 ppm |
| | | C2 = B − C1 |

Applicability conditions:

- SOx: installation of pollutant reduction equipment or reduction of SOx by improving production process, in addition to emission concentration of 100 ppm or below.
- NOx: installation of pollutant reduction equipment or reduction of NOx by improving production process, in addition to emission concentration below 50% of NOx emission standard announced on the day of implementation by agencies under the authority of the central government or 100 ppm or below.
- Installation of pollutant reduction equipment or prioritized adoption of Optimal Reduction Technology for Fixed Pollution Sources in improving production process.

Source: created with data from TEPA (2008).

petroleum used by the pollution sources as the standard. The first amendment included in 1997 stated that fees were collected based on the amount of gasoline (fuel) used, or the amount of emissions, and the concentration of pollutants from automobiles. It was also made possible to collect fees from the users of gasoline (buyers) or the producers or importers based on the sales. There was no change associated with the non-point sources in the second amendment. A change was made in the third amendment, which was approved in 1999: fees would be collected from retailers or importers based on the kind and quantity of air pollutants emitted, or based on the kind, components, and amount of fuel.

However, even though fees were supposed to be collected from all users of gasoline as of April 1999, in reality they were collected through the retailers of gasoline with a fee included in the fuel price. A change was made to the law so that the fees would be collected from the retailers or importers of fuel. Thus, the gasoline suppliers were obliged to pay the fees.

### Destination of the APC fees

Financial institutions that serve as agents for fee collection send the collected fees directly to the central government and a regional government. More specifically, 40 percent of them go to the central government's APC fund; 60 percent, to the APC fund set up by a municipality, county, or city government. The regional environmental protection bureaus support the collection effort by reminding non-payers, or those who have made an insufficient payment, of the required payment or giving them additional reference documents, in addition to examining and confirming the collection of the APC fees.

### Collection of the APC fees to date

The total amount of APC fees collected in the first phase according to the quantity of consumed fuel containing SOx as well as the actual content of sulfur in the fuel was 8.69 billion yuan between July 1995 and June 1998. The amount associated with coal and fuel oil was 6.54 and 35.1 billion yuan, respectively. According to data on the collection of the SOx and NOx APC fees for various seasons, approximately 5,400 entities nationwide paid the fees. The total amount collected by June 2008 was approximately 17.43 billion yuan, with an average of 440 million yuan collected per season. An examination of changes in the amount of the collected APC fees reveals that the main reason for the decrease in the amount over the years is an obvious decline in emissions of pollutants. The share of the non-point and point sources in the APC fund changed from 24.0 and 76.0 percent in 1997 to 68.7 and 31.3 percent in 2006, respectively. The emissions of SOx and NOx fell by 42.6 and 25.5 percent, respectively, compared to their levels when the fee system first started. This provides an explanation as to why the collected amount tended to gradually decline. As the effect of the APC fees spread, many polluters in both private and public sectors installed pollutant reducing equipment or used low-pollution fuel, which led to reduced pollutant emissions. Because of the APC fees, the annual average concentration of $SO_2$ in the atmosphere fell from 9 ppm in 1994 to 4 ppm in 2004.

In addition, data on the collection of the VOCs APC fee in the third phase show that approximately 4,500 entities nationwide paid it. The total amount collected by July 2008 was 695 million yuan, with a seasonal average of 140 million yuan.

## Effects of the APC fees

### Environment budget/expenditures

Taiwan's environmental budget/expenditure's scale increased from NT$15.2 to 45.4 billion between 1989 and 2006, with an annual average growth rate of 11.1 percent over the 18 years. Also, the growth rate for the subcategories—general administration, overall planning, research and development, air pollution, noise, water quality protection, waste, sanitation and toxic materials, and others—was 34.2, 2.6, 1.9, 16.9, 18.7, 86.7, 28.4, 19.0, and –3.1 percent, respectively. The budget/expenditures for water quality protection drastically increased. The share of waste disposal was the highest at around 70 percent in recent years—a dramatic rise from 30–40 percent in the 1990s. The budget/expenditures for combating air pollution and noise came from the same category from 1989 to 2000. The rapid expansion of the budget from 1996 to 1998 was the result of collecting the SOx APC fee (July 1, 1995–) and the fee for particulate pollution from building constructions (July 1, 1997–) which were temporary fiscal measures effective until their transfer to the APC fund in 2001.

### Air pollution control policy

The Pollution Standards Index (PSI) is often used to measure air quality. The PSI per day is obtained by taking the maximum of the subindexes of the measurements of the concentration of air pollutants for the day such as particulate matter (PM10), sulfur dioxide ($SO_2$), nitrogen dioxide ($NO_2$), carbon monoxide (CO), and ozone ($O_3$). The categories for air quality are determined as follows: "good" ($0 < PSI < 50$), "normal" ($51 < PSI < 100$), "bad" ($101 < PSI < 199$), "very bad" ($200 < PSI < 299$), and "harmful" ($299 < PSI$).

There were 258 air quality observation centers in Taiwan as of 2006. Among them, 76 are automated centers directly managed by the central government agency (TEPA), 14 are automated centers managed by regional (provincial or city) environmental protection bureaus, and 168 are staffed observation centers (TEPA 2007). The proportion of PSI categories "bad" and "very bad" during 23 years from 1984 to 2006 was around 15 percent for the 1984–1991 period, but it fell to 4.16 percent in 2006, showing a gradual tendency for improvement. Especially, the improvement made between 1991 and 1996 was substantial.

Comparing the effective amount of air pollutant emissions for 1988, 1995, and 2003, among major pollutants, the emissions of total suspended particulate

(TSP) matter, sulfur oxide (SOx), and carbon monoxide (CO) decreased by 48.6, 48.8, and 21.2 percent, respectively, between 1988 and 1995. Consequently, the proportion of the PSI exceeding 100 fell. The decreased emission of pollutants is ascribed to progress in emissions regulations. Also, while TSP and SOx decreased by 50.8 and 57.5 percent, respectively, between 1995 and 2004, non-methane hydrocarbon (NMHC), CO, and lead (Pb) increased by 51.2, 17.2, and 9.7 percent, respectively. The increase in these air pollutants is deemed to be a result of a rise in the number of automobiles and their size (Taiwan Environmental Protection Administration, 1990, 2000, 2002, 2004, 2007).

A plan to police point sources jointly from the sky and ground was put into action starting in September 1989. Air policemen and TEPA investigators on helicopters searched for pollution sources, and TEPA investigators on the ground executed policing operations. As a result, a total of 8,778 factories emitting pollutants illegally or engaging in outdoor burning had been found by June 1995, and 648 (7.4 percent) of them were charged with violation. Also, the inspection of flues at point sources began on April 12, 1990. By the end of the operation at the end of June 1995, 3,164 factory flues were examined, and 581 (18.4 percent) did not pass the inspection.

Those who did not pass the inspection were punished with fines or formal charges filed against their violation and, simultaneously, were ordered to make improvements within a specific time frame. The series of policing efforts, as well as regulations on emissions, contributed to the reduction of air pollution. However, NOx emissions, which were mainly from automobiles, did not show a tendency to decline. Furthermore, as mentioned earlier, based on the APC Act that was enacted on March 23, 1995, the collection of APC fees began on July 1. There is controversy as to the effect of this fee and the use of the revenue from it, but it is certain that the Taiwanese government is proactively using economic instruments as measures to incorporate environmental considerations into the market framework.

## Conclusion

With the collection of the APC fees, Taiwan intends to give firms and individuals incentives to make efforts to prevent pollution or reduce pollutant emissions. The findings of this chapter on the result of the APC fees can be summarized as follows.

First, they provide incentives to switch fuels. In the first phase of the APC fee system, fees were collected based on the amount and characteristic of fuel consumed. Since fees were high for fuels with high concentration of sulfur, after the collection of the fees started many firms and individuals reduced the use of petroleum coke, which contains a high level of sulfur, and began proactively purchasing coal, which is low in sulfur content. This incentive for fuel substitution contributed to the decline of average sulfur content from 1.2 to 0.6 percent. Moreover, the implementation of an economic instrument, which waived the APC fees if clean energy (e.g., natural gas, liquefied petroleum gas) was used, indirectly provided firms and individuals incentives to shift to the use of clean energy, contributing to a certain degree of pollution reduction.

Second, the APC fees provided an incentive to install pollutant-reducing equipment. During the period from July 1995 to June 1998, more than 20 companies in electric power, cement, and large-scale boiler industries, as well as firms and individuals using coal and fuel oil, applied for a discount or waiver. Prior to the implementation of the discount-waiver program, according to statistics provided by TEPA, many firms and individuals had had their SOx emissions just below levels specified by emission standards and had not attempted to reduce emitted pollutants efficiently through sufficient use of pollution control equipment. However, after the program began, they tried to receive a high level of discount or a waiver by reducing their SOx emissions. They did so through purchasing coal, which is low in sulfur content, installing equipment for reducing SOx pollution, or operating pollution controlling equipment more carefully.[6] The second phase of the APC fee collection started in July 1998. In the beginning, the fees were calculated and collected based on the actual amount of pollutant emissions. However, in order to give firms and individuals incentives to proactively reduce emissions of pollutants and improve air quality, preferential rates were applied to those who installed pollution controlling equipment, those who improved production processes and reduced SOx and NOx emissions more effectively in those processes, and those who kept their emissions low relative to emission standards. As a result, firms and individuals began proactively installing pollution control equipment, which reflects the effect of incentives for reducing SOx and NOx emissions.

## Notes

1  In May 1990, the TEPA commissioned a university to conduct a study entitled "Research Plan for Details on the Implementation of the Air Pollution Control Act and Methods to Collect Air Pollution Control Fees," which was completed in the following year.

2  For example, in obtaining the base charge rate for non-point sources, a cargo tax and a pollution tax levied on the use of fuel are used as benchmarks.

3  In the Legislative Yuan session on May 31, 1995, the budget for the APC fees for FY1996 was passed, but its amount was reduced to 6.9 billion yuan from 9.9 billion yuan. During the deliberation on the budget for FY1997, the amount that was the same as the previous year's (6.9 billion yuan) was submitted for the APC fees. However, some environmental organizations and legislative members argued that it should be removed since the implementation of the budget for the APC fees for FY1996 was inappropriate. As a result, on May 1, 1996, a joint session between the Internal Administration Committee and Budget Committee of the Legislative Yuan removed the entire budget. However, after several rounds of discussions, the budget for the APC fees for FY1997 was passed in the end with a reduced amount of 4.7 billion yuan.

4  In the first stage, a deputy minister of TEPA, accompanied by the chief of its Department of Air Quality Protection, visited 33 legislative members and provided explanations. In the second stage, information was extensively given to the legislative members and parliamentary office staff members. In 1995, during the period from March 23 through March 28, TEPA again visited all 145 legislators, answered all questions that

they had, and provided additional documentation. These efforts were made to have the legislators understand further the content of the policy and gain their support.

5 On May 9, 1997, during the 1,070th session of the Judicial Yuan, the Council of Grand Justices issued an interpretive document (Release No. 426, 1997), responding to the request for Constitutional interpretation submitted on June 12 by 67 legislative members including Legislator H. According to the Council's interpretation, the method used to collect the APC fees did not have any conflict with the assignment of authority prescribed by the APC Act. However, since the treatment of "other pollutants" discharged by the point sources violated equity under the law, the Council suggested that fees be levied based on the sources of pollutants.

6 For instance, the chemical factories of seven companies including Companies F and N installed, one after another, a total of ten flue gas desulfurization (FGD) systems. This led to a reduction in the SOx content of emissions from 350 ppm to 30 ppm. The discount-waiver program reduced SOx emissions by 33,000 tons/year (TEPA 2008).

# References

Directorate General of Budget, Accounting and Statistics, Executive Yuan (2007) *Result of Devising Green GDP, 2007* (in Chinese Big 5).

Ministry of Finance, Executive Yuan (2008) *Annual Fiscal Statistics Report, 2008* (in Chinese Big 5).

Murakami, R. (2006) "Taiwan's Recycling and Industrial Waste Policy." Report on the Information Dissemination Project on Recycling and Industrial Waste Policy of Asian Countries, Institute of Developing Economies, Japan External Trade Organization (in Japanese).

Nanbu, K. (2007) "Recycling in Taiwan: The Current Situation and Issues," *Journal of the Ohara Institute for Social Research*. No. 580 (March), pp. 11–16 (in Japanese).

OECD (2001) *Environmentally Related Taxes in OECD Countries: Issues and Strategies*, Paris: OECD.

Pearce, D.A., A. Markandya, and E.B. Barbier (1989) *Blueprint for a Green Economy*, London: Earthscan.

Taiwan Environmental Protection Administration (TEPA), Executive Yuan (1990) *Taiwan Environmental Information* (in Chinese Big 5).

Taiwan Environmental Protection Administration (TEPA), Executive Yuan (2000) *Taiwan Environmental Information* (in Chinese Big 5).

Taiwan Environmental Protection Administration (TEPA), Executive Yuan (2002) *Water Pollution Control Act* (in Chinese Big 5).

Taiwan Environmental Protection Administration (TEPA), Executive Yuan (2002) *Basic Environment Act* (in Chinese Big 5).

Taiwan Environmental Protection Administration (TEPA), Executive Yuan (2004) *Forum on Water Pollution Control in River Basins and Water Recycling* (in Chinese Big 5).

Taiwan Environmental Protection Administration (TEPA), Executive Yuan (2007) *Annual Statistical Report on Environmental Protection in Taiwan* (in Chinese Big 5).

Taiwan Environmental Protection Administration (TEPA), Executive Yuan (2008) *Annual Statistical Report on Environmental Protection in Taiwan* (in Chinese Big 5).

Uzawa, H. (1995) *Economics of Global Warming*, Tokyo: Iwanami Shoten (in Japanese).

von Weizsaeker, E.U. (1990) *Erdpolitik: Oekologishe Realpolitik an der Schwelle zum Jahrhundert der Umwelt*, Darmstadt: Wissenschaftliche Buchgesellschaft.

# 5 Greening the government budget in Vietnam

## Recent changes and future challenges

*Le Thi Kim Oanh and Akihisa Mori*

### Introduction

The government of Vietnam has developed environmental institutions to address worsening environmental pollution. It split the Ministry of Science and Technology and Environment (MOSTE) into the Ministry of Natural Resources and Environment (MONRE) and the Ministry of Science and Technology (MOST), while the Vietnam Environmental Protection Agency (VEPA), together with the provincial environmental management offices, the Departments of Natural Resources and Environment (DONREs), were assigned to MONRE. Other industrial environmental management units were established in the Ministries of Industry, of Agriculture and Rural Development, and of Fishery. Various ministries, including Health, Construction, Transportation, Industry, Fishery, Agricultural and Rural Development issued regulations and action plans on environmental protection for their sectors. People's Committees of provinces and centrally affiliated cities issued strategies, planning and action plans on environmental protection within their jurisdiction, in accordance with the guidelines of the central government.

The government has also issued environmental legislation and policy measures. It published the "National Strategy for Environmental Protection to 2010 and vision toward 2020" in 2003, of which cleaner production was posited as one of the main activities. This strategy led the way to the revision of the Law on Environmental Protection (LEP) in 2005, which sets out the requirements for environmental impact assessment (EIA), implements strategic environmental assessment (SEA) and describes the extended producer's responsibility in waste management. The government also published the Decision No. 64/2003 on environmental fees for wastewater to make stringent regulations on companies that cause serious environmental pollution. It decided to force polluting companies to relocate, close down or clean up effluents by 2007. Ho Chi Minh City (HCMC) also released decision No. 80/2002/Q-UB on the approval of the "Relocation of Polluting Enterprises to Industrial Zone Program," for the purpose of reducing industrial pollution, promoting technological renewal and implementing spatial separation between industrial and residential area (Khoa and Ho, 2006).

The government has also developed environmental fiscal mechanisms. It published Decision No. 67/2003/ND-CP authorizing the government to impose charges on industrial and domestic polluters and to allocate them to the Vietnam Environmental Protection Fund (VEPF) and the DONRE. It also issued Directive No. 41-NQ/TW in 2006 planning to annually allocate at least 1 percent of the state budget expenditure for environmental protection activities from 2007. This portion would be increased in the following years depending on the financial resource capacity as well as the demand for environmental investment. This directive is applied not only for the central government, but also for local governments. In response, some provinces and centrally affiliated cities established a local environmental protection fund to secure the portion.

Effectiveness of these enhanced environmental legislation and policy measures has been limited so far. The relocation project at HCMC has not gone well due to incomplete preparation in terms of the realization of infrastructure, the construction of relocation sites, the establishment of support mechanisms, and sufficient commitment from businesses (Khoa and Ho, 2006). Cleaner production has been gradually disseminated and incited some pulp and paper mills to recognize the environmental and financial benefits, but they have not provided enough incentives to comply with environmental standards (Luken and Van Rompaey, 2007). Increased prices for water and energy supplies, enforcement of environmental regulations and pressure from local communities are required for further diffusion. Current EIA legislation has a strong legal basis, but has shortcomings in practice relating to public participation in terms of method, timing of consultation and impacts on decision-making (Clausen *et al.*, 2011).

Effectiveness of the environmental fiscal mechanism, however, has not been explored so far in Vietnam. Referring to experiences in China, a proportional levy rate with inflation adjustment, strict enforcement of environmental regulations and efficient revolving mechanism are essential for environmental financing through pollution levy to work effectively for emission reduction.

This chapter describes the evolution of environmental fiscal mechanism in Vietnam both at central and local government levels and analyzes the effectiveness of local environmental fiscal mechanism. We will pay special attention to the two big and province-like status cities of HCMC and Danang city, and Binh Duong Province that is located next to HCMC and has 28 industrial parks with over 1,200 industrial enterprises, thus faces the challenges of enhanced industrial pollution.

## Integration of environmental concerns in the state budget planning

Before the establishment of the MONRE, environmental protection function at the central level was a part of governmental function responsible by MOSTE and its National Environmental Agency. Due to this institutional structure, investment in environmental protection from the state budget accounts for approximately 150–200 billion VND each year, out of which 75 percent is allocated for projects on basic investigation, infrastructure improvement, and environmental protection, with the remaining 25 percent being considered as development investment budget for projects on environmental monitoring and analytical stations for ministries and localities. The annual budget for environmental protection is taken from the item of recurrent expenses for science and technology in the state budget. Some capital sources for environment-related activities have been allocated from the category of economic development expenses.

Official development assistance (ODA) capital for the environment seems to increase annually but it accounts for a fairly small percentage of total ODA (Ministry of Natural Resources and Environment, 2005). According to MONRE, total ODA in the period from 2000–06 was about 1,465.35 billion VND, of which 663.35 billion VND were soft loans, and 802 billion VND were aid packages.

The establishment of the MONRE in 2002 has shown the strong effort of the government toward environmental protection in Vietnam. However, at that time, the government investment in the environment has been limited as the portion of the state budget allocated to environmental management and protection at both central and local levels has been low and inadequate. Financial resources invested in the environment have not yet been defined as a separate category of expenditure in the state budget. This causes some difficulties not only for managing resources in a consistent manner, but also for providing the basis for calculating the share of environment in the total investment and for evaluating efficiency and effectiveness of environmental protection efforts.

In 2006, in order to have necessarily effective measures for implementation of Directive No. 41-NQ/TW of the Politburo of Vietnam Communist Party on Environmental Protection and the approved National Strategy for Environmental Protection until the year 2010 and vision toward 2020, the National Assembly approved that from 2007, the government will allocate at least 1 percent of the state budget expenditure for environmental protection activities annually, with gradual increases foreseen in the following years depending on the budget financial resource capacity as well as the demand for environmental investment. This important decision was also considered as a great effort and strong government commitment in dealing with environmental pollution and supporting environment protection of the country (Table 5.1).

The five-year plan of 2006–10 was seen as the first planning period of the countries in which economic, social and environment concerns are equally integrated in the state social-economic development plan. In this plan, there were eight environmental indexes such as forest coverage rate, number of industrial pollution hot spots being eliminated and urban solid waste collection rate.

Thanks to this decision of the National Assembly, the expenditure for the environment has been increasing year by year since 2006. This increment is very supportive to meet the increasing demand on environmental protection activities needed to be taken by the government authority at all levels. The environmental expenditure in the state budget was 2.9 trillion VND in 2006. Also, the

Table 5.1 The state budget, budget expenditure and environmental expenditure from 2000 to 2010 (in billion VND)

| Year | 2000 | 2002 | 2003 | 2004 | 2005 | 2006 | 2007 | 2008* | 2009* | 2010* |
|---|---|---|---|---|---|---|---|---|---|---|
| GDP | – | 535,762 | 613,443 | 715,307 | 839,211 | 973,791 | 1,144,014 | 1,478,695 | 1,813,000 | 1,931,300 |
| Total state budget revenue | 90,749 | 123,860 | 158,056 | 198,614 | 238,686 | 245,900 | 300,900 | 332,080 | 404,000 | 462,500 |
| Total state budget expenditure | 108,961 | 148,208 | 178,541 | 221,792 | 279,873 | 294,400 | 357,400 | 398,980 | 491,300 | 582,200 |
| Of which | | | | | | | | | | |
| Science, technology and environmental expenditures | 1,243 | 1,852 | 1,853 | 2,362 | 2,584 | 2,900 | 3,500 | 3,883 | 5,150 | 6,230 |
| Central government budget | – | – | – | – | – | 500 | 530 | 580 | 850 | 980 |
| Local government budget | – | – | – | – | – | 2,400 | 2,970 | 3,303 | 4,300 | 5,250 |
| % in total budget | 0.011 | 0.012 | 0.010 | 0.011 | 0.009 | 0.010 | 0.010 | 0.010 | 0.010 | 0.011 |
| % in GDP | – | 0.0035 | 0.0030 | 0.0033 | 0.0031 | 0.0030 | 0.0031 | 0.0026 | 0.0028 | 0.0032 |
| Expenses for national targeted programs | – | 4,575 | 5,161 | 5,572 | 6,852 | 8,631 | 9,175 | 10,382 | 13,452 | 15,984 |
| Environmental related programs | – | 569 | 662 | 766 | 804 | 963 | 1,170 | 1,464 | 1,975 | 2,652 |
| Of which | | | | | | | | | | |
| Reforestation | 293 | 350 | 392 | 500 | 503 | 610 | 720 | 820 | 1,000 | 1,425 |
| Safe water and sanitation | – | 219 | 270 | 266 | 301 | 353 | 430 | 622 | 935 | 1,099 |
| Energy saving | – | – | – | – | – | – | 20 | 22 | 40 | 61 |
| Climate change | – | – | – | – | – | – | – | – | – | 68 |

Data source: collected from annual state budget reports, provided by Ministry of Finance, Vietnam.

Note
* Budget planning.

expenditure for environmental protection has been clearly defined as a separate item in the sectoral budget expenditure. Furthermore, the government changed the budget allocation between central and local governments so that the latter would gain more financial sources for environmental protection activities.

The state budget expenditure planning for 2007 was 3.5 trillion VND that increased 20.7 percent compared to the 2006 budget, and occupied over 1 percent of the state budget expenditures. This amount was decided in order to ensure financing sources for implementing the important environment tasks regulated by legal documents such as the Law on Environmental Protection, the Resolution No. 41/NQ issued by the Political Bureau of the Vietnam Communist Party, and implementing the environmental protection activities urged by the Prime Minister Decisions such as Decision No. 34/2005/QĐ-TTg on the Government Action Plan for Environmental Protection, Decision No. 256/2003/QĐ-TTg on the approval of National Environmental Protection Strategy up to 2010 and the vision toward 2020, and Decision No. 328/2005/QĐ-TTg dated December 12, 2005 on the Approval of National Plan for Environmental Pollution Control up to 2010.

The state budget expenditures were increased during 2008–10 to meet additional demands for the National Targeted Program[1] on Clean Water Supply, Sanitary for Rural areas and Coping with Climate Change, as well as various focused environmental protection activities and programs (Table 5.1). An additional 300 billion VND were supplied to the charter capital of the National Environmental Fund to meet the objective of increasing VEPF capital up to 800 billion VND until 2010.

The VEPF was established in 2002 and assigned for the environment, such as wastewater exhaust gas, solid waste and mineral exploitation. Until the end of 2010, there were about 120 environmental projects which received soft loans or grants from VEPF with the total disbursement about 438 billion VND. At the same time, total capital of VEPF, which was established with the state start-up amount of 200 billion VND, reached 900 billion. However, the complicated and lengthy procedure has prevented firms from accessing this fund.

The environmental protection expenditure from the state budget is around 0.3–0.4 percent of the GDP. However, it can be higher if national targeted programs are included (Table 5.2).

Also in the five-year plan 2006–10, Vietnam has received over 60 international cooperation projects with a total funding amount from ODA of US$223 million.

The government also firmly committed to allocate 1 percent of the government budget expenditures annually for environmental protection acknowledging the huge financial demands for environmental protection at present. For example, the amount of capital for the project on environmental improvement and pollution treatment at craft villages was estimated at about 17,678 billion VND per year. Recent National Assembly meetings received a numbers of proposals to increase state budget expenditures for environmental activities up to 2 percent per year.

Table 5.2 Total budgets of line ministries relating to environmental protection (in million VND)

| Ministries | 2006 | | 2008 | | 2009 | | 2010 (planning) | |
| --- | --- | --- | --- | --- | --- | --- | --- | --- |
| | Total expenditure | Environmental protection expenditure | Total expenditure | Environmental protection expenditure | Total expenditure | Environmental protection expenditure | Total expenditure | Environmental protection expenditure |
| Agricultural and Rural Development | 3,153,848 | 8,100 | 3,589,162 | 22,000 | 5,714,417 | 25,620 | 6,172,712 | 17,895 |
| Transportation | 8,944,175 | 5,500 | 8,509,123 | 12,990 | 8,844,539 | 7,820 | 10,327,159 | 12,416 |
| Industry and Trade | | 16,100 | 1,137,105 | 18,250 | 1,387,760 | 19,400 | 1,644,003 | 18,860 |
| Construction | 1,848,410 | 10,000 | 807,356 | 9,420 | 884,027 | 9,500 | 1,437,291 | 8,800 |
| Health | 2,437,841 | 11,000 | 3,955,586 | 34,110 | 4,622,569 | 40,641 | 5,498,341 | 36,870 |
| Education and Training | 2,750,236 | 4,200 | 4,057,535 | 13,950 | 3,868,160 | 13,550 | 4,361,017 | 9,500 |
| Science and Technology | 402,915 | 2,700 | 579,839 | 3,000 | 1,011,815 | 5,430 | 933,120 | 5,500 |
| Culture, Sports and Tourism | 1,000,840 | 4,000 | 1,567,995 | 6,180 | 1,813,450 | 13,850 | 1,984,432 | 12,006 |
| Natural Resource and Environment | 880,479 | 112,000 | 1,473,931 | 150,000 | 1,641,443 | 165,000 | 2,073,650 | 180,000 |

Data source: collected from annual state budget reports, provided by Ministry of Finance, Vietnam.

## Evolution of the environmental protection budget and policy implementation at the local level

### *Ho Chi Minh City*

#### *Administrative structure for environment management*

The local environmental authority in HCMC is designated to the Department of Natural Resource and Environment which was established in 2003 as a result of the renewal of the environmental management structure at all levels decided by the government. The organization structure of HCMC DONRE is a bit different from the common administrative structure in other province/cities. DONRE of HCMC has both an Environmental Management Division and Environmental Protection Agency at the same time. The function of solid waste management is also specialized for the division of Solid Waste Management while in most other provincial DONREs, solid waste management function is still being executed by the Departments of Transportation and urban environmental companies under their authority. This unique organizational structure was reorganized to unify the Division of Environmental Management and the Environmental Protection Agency in 2011.

#### *Local regulations on environmental protection*

Most of the current legislative documents on environmental management at the city level are decisions issued by the City's People Committee in order to specify in details and to implement general Law and Decrees stipulated by the government. Some important ones are:

- Decision No. 99/2006 QD-UBND dated July 11, 2006 on issuing Plan to implement Action Plan of the Political Bureau of the City Communist Party in accordance with Resolution No. 41-NQ/TW on Environmental Protection in the period of industrializing and modernizing by the Vietnam Communist Party.
- Decision No. 190/2004 and 139/2007 of the HCMC People Committee on collecting the environmental protection fees for wastewater in Ho Chi Minh City.
- Decision No. 103/2004/QD-UB on issuing the Plan on Solid Waste Management in Ho Chi Minh City.
- Decision No. 4329/QD-UBND to approve the Plan for Implementation of the Dong Nai River Basin Environmental Protection Project.

Since 2004, HCMC DONRE has been preparing a master plan for environmental protection of the City. However, after many adjustments, this master plan has yet to be approved. The main reason is that rapid growth of the economy in HCMC seems to cause difficulties for policy makers on environmental management in making such a master plan to meet the requirements of society.

At present, DONRE is still working on the preparation of a new proposal on a master plan for environmental management in HCMC until 2020.

## Local government budget and environmental fiscal expenditures

Together with Hanoi capital, Ho Chi Minh City has an annual city government budget which is the biggest in the country. Since 2006, the HCMC government has committed to follow the central government decision of saving at least 1 percent of the city government budgets for environmental protection activities (Table 5.3).

The governmental budget for environmental expenditure has been steadily rising during the last years, especially since 2006 when the central government of Vietnam had a strong commitment on allocating 1 percent government budget for the environment. Since then, there have been many environmental programs initiated by the local government to meet the demand of environmental protection of the local citizens and to ensure sustainable development (Table 5.4).

## HCMC waste recycling fund

Recycling activities in HCMC are implemented as a flourishing business. However, as with other cities of Vietnam, this sector is being developed informally in this city. To promote the development of the recycling industry, HCMC People's Committee issued Decision No. 5488 dated November 30, 2006 to establish the HCMC Waste Recycling Fund (REFU).

REFU is a non-profit financial organization under management of HCMC's DONRE. Its function is to supply loans to the programs, projects and conduct

*Table 5.3* Budget expenditure and environmental expenditures in HCMC (in million VND)

|  | 2008 | 2009 | 2010 |
|---|---|---|---|
| Total budget | 30,572 | 30,170 | 29,524 |
| Environmental protection expenditure | 776 | 1,371 | 1,487 |

Source: Department of Finance, HCMC.

*Table 5.4* Budget expenditure at the HCMC environmental protection agency (in million VND)

|  | 2005 | 2006 | 2007 | 2008 | 2009 | 2010 | 2011 |
|---|---|---|---|---|---|---|---|
| Total budget | 8,094 | 9,277 | 13,930 | 17,632 | 14,639 | 16,071 | 21,721 |
| Professional activities | 5,276 | 5,732 | 9,463 | 9,965 | 9,943 | 12,005 | 13,021 |

Source: HCMC Environmental Protection Agency.

studies in the field of waste management, recycling and reduction. All the activities/operations of REFU are focused on the promotion of recycling so that the fund is also considered as a kind of environmental fund.

The main challenge of this recycling fund is lack of capital. Since 2006, this fund has received a 20 percent share of the charter capital from the budget of the city. However, REFU has organized many activities to promote the development of recycling activities in this city. Not only focusing on the recycling enterprises, this fund also conducts other activities to improve the public awareness on waste management in general and recycling in particular.

### Binh Duong Province

*Administrative structure for environment management in Binh Duong Province*

Within Binh Duong's DONRE, the functional unit responsible for environmental issues is the Environmental Protection Agency which was previously known as the Environmental Management Division.

As a technical support institution, the Center of Environmental Monitoring was established to conduct more operational activities of environmental quality monitoring, environmental supervision, and providing advisory service on development and application of environmental technology to environmental organization and enterprises.

In 2009, an environmental protection fund was created. The fund is running as a functional division under DONRE management.

*Local regulations on environmental protection*

Before 2010, there were not many regulations which were made at the local level for environmental management and environmental protection at Binh Duong Province. Since 2010, Binh Duong People Committee has intensively issued many decisions, and urged on the preparation of a master plan for environmental protection in this province.

On December 10, 2010, the People Council of Binh Duong Province agreed on a resolution to issue a master plan on environmental protection in Binh Duong for the period 2011–15, and the People Committee approved and issued it in 2011.

Following the approval of this plan, the DONRE is preparing a draft on Environmental Protection Strategy for Binh Duong toward 2020 with the vision for 2030. In this strategy, a number of ambitious environmental objectives have been planed such as increasing the use of renewable energy to 5 percent, at least 50 percent of industrial enterprises applying pollution prevention and cleaner production measures, 100 percent urban residents are able to access clean water supplying in 2020.

*Local government budget and environmental fiscal expenditures*

Similar to the case of HCMC, the local environmental budget expenditures have been steadily rising since 2006. The amount of environmental expenditures soared in 2009 and 2010 (Table 5.5). This increase accrued to the central government decision to foster environmental protection in fiscal policy.

*Environmental protection fund of Binh Duong Province*

Following the start of the Recycling Fund in Ho Chi Minh City, Binh Duong is the second locality in Vietnam to initiate a local environmental protection fund (EPF) as a supportive financing source for environmental protection activities. Binh Duong EPF was established by Decision No. 805/QĐ-UBND of the Binh Duong People Committee on March 5, 2009 and officially went into operation from May 2009.

At the beginning, the charter capital of Binh Duong EPF was 20 billion VND financed by local government budget. In 2010, the EPF capital was raised to 31.5 billion VND, of which ten billion VND came from the local government budget and 1.5 billion VND was added from the environmental protection fee revenue. It also means that, in Binh Duong, the revenue from the environmental protection fee on industrial wastewater has been allocated for common environmental protection activities through the operation of the EPF whereby the environmental protection fee revenue is a source of capital.

*Table 5.5* Budget expenditures and environmental expenditures in Binh Duong Province during 2006–10 (in million VND)

|  | *2006* | *2007* | *2008* | *2009* | *2010* |
|---|---|---|---|---|---|
| Total province budget | 2,774,066 | 3,023,148 | 3,887,671 | 6,202,326 | 7,500,000 |
| Environmental expenses | 40,996 | 45,317 | 46,896 | 74,500 | 107,919 |
| at provincial level | 994 | 1,885 | 2,717 | 6,517 | 8,426 |
| at district level | 40,002 | 43,432 | 44,179 | 67,983 | 99,493 |
| Share of environmental expenses | 1.48 | 1.50 | 1.21 | 1.20 | 1.44 |

Source: Binh Duong Natural Resource and Environment, Binh Duong DONRE.

*Table 5.6* Environmental expenses in Binh Duong DONRE budget (in million VND)

|  | *2007* | *2008* | *2009* | *2010* |
|---|---|---|---|---|
| Environmental monitoring | 482,775 | 376,155 | 2,606,913 | 3,358,417 |
| Other expenses | 1,401,844 | 1,159,522 | 3,910,369 | 4,083,057 |

Source: Binh Duong Natural Resource and Environment, Binh Duong DONRE.

Binh Duong EPF offers soft loans to the environmental protection projects proposed by industrial enterprises such as wastewater, air and solid waste treatment system, manufacturing of environmental friendly products, products which are produced from recycling materials, and projects on raising environmental awareness. The loan maturity date is set at a five-year maximum.

At the end of 2010, after over one year of operation, the environmental protection fund of Binh Duong had given out loans for 12 projects with the total loan amount of 20 billion VND. Among those 12 projects, there were seven projects on wastewater treatment and five projects related to solid waste treatment.

### Danang City

*Administrative structure for environmental management in Danang*

Generally speaking, environmental management in Danang city has its own administrative structure which is very typical for most provinces in Vietnam. The function of environmental protection is concentrated at the Environmental Protection Agency under the DONRE. Besides EPA, there is a center for conducting most technical activities which serve for environmental management, such as environmental monitoring, environmental (air, water, solid) sampling and analysis, and environmental technology consultancy.

Different from other cities, since February 2010, Danang City People Committee has decided to establish Danang Drainage and Wastewater Treatment Company under the administration of Danang DONRE. This is one step to improve the control of the wastewater problem in the city after four wastewater treatment plants have been completed and now they need to be operating in a proper manner, and the revenue collected from environmental protection fees on wastewater needs to be used efficiently to facilitate wastewater pollution control.

*Local regulations on environmental protection*

Several regulations related to environmental protection and management at the local level have been initiated in Danang since 1997, such as Decision No. 3903/QD-UB dated October 22, 1997 of Danang People Committee on issuing regulations on solid waste management in the city. At the end of 2000, Danang People Committee issued Decision No. 142/2000/QD-UB dated December 29, 2000 on general activities for environment protection.

However, Danang has not had a master plan or a strategy for environmental management yet. In 2009, a JICA-supported study on Integrated Development Strategy for Danang city and its neighboring area was implemented in Danang to publish an overall plan on environmental management for the whole city up to 2020. The report is expected to be submitted to the city government for approval.

In addition, the plan of "Developing Danang – the environmental city" authorized by Danang People Committee shows some targets on environment quality improvement for Danang city, for example, until 2020 collection ratio should be more than 90 percent, domestic wastewater treatment and industrial wastewater treatment should be 100 percent, and 25 percent of wastewater and 70 percent of solid waste are recycled.

*Local government budget and environmental fiscal expenditures*

Similar to other localities in Vietnam, before the central government decision on the change of state budget expenditures toward facilitating environmental protection in 2006, the local government budget of Danang city had no separate item for environmental activities.

From 2007 to present, the environmental expenses are always allocated clearly in the city government budget every year (Table 5.7). Although the share of environmental expenses in the city's total budget expenditure has not been fully following the general decision of the central government (i.e., leaving at least 1 percent of budget expenditure to environmental expenses), it seems to strictly comply with the financial objectives set out in the Plan for "Developing Danang – The Environmental City" that was initiated by the city government in 2008. In this exciting "green" plan, it was stated that the city government will save 0.8 percent of the city budget for environment in 2010, and this portion will continue to increase to 1.5 percent in 2015.

Environmental financing sources from the city government are mainly granted to DONRE and the Urban Environmental Company (URENCO) (Table 5.8). The funding source for environmental management of the responsible authority (DONRE) has been increasing rapidly during the period 2007–11. Budget expenses which are allocated to URENCO are used mainly for urban environment and sanitation.

### *Highlighted features of environmental expenditures at three cities/ provinces*

Looking at the environmental expenditures at the local level of three cities/ provinces described above, there are some features which are similar among those localities. In accordance with the central government order, all three localities have gradually increased their environmental expenditures and reached 1 percent of their annual expenditure. This shows strong government commitments to environmental protection not only at the central level but also at the local level. They allocate environmental protection expenditure to a variety of fields, while the lion's share goes to wastewater and solid waste management. The rest covers general environmental administration and management, education, training and information relating to the environment.

The differences between provinces/cities, to some degree, reflect the evolution of decentralization in state management in Vietnam since economic reform has been instigated. Through decentralization, power and responsibility are transferred from the central level to local governments, thereby bringing

*Table 5.7* Environmental fiscal expenses and its shares in total local budget expenditures of Danang city (in million VND)

| | 2007 | 2008 | 2009 | 2010 | 2011 |
|---|---|---|---|---|---|
| Total budget expenditure | 5,347,939 | 8,185,833 | 9,133,065 | 6,330,152 | 8,563,581 |
| Environmental expenses | 35,038 | 40,107 | 44,617 | 50,886 | 100,332 |
| Share of environment expenses in total budget | 0.66 | 0.49 | 0.49 | 0.80 | 1.17 |

Source: Department of Finance, Danang City.

*Table 5.8* Sharing environmental expenses between DONRE and other city departments (in million VND)

|  | *2007* | *2008* | *2009* | *2010* | *2011* |
|---|---|---|---|---|---|
| Environmental expenses | 35,038 | 40,107 | 44,617 | 50,886 | 100,332 |
| DONRE | 2,804 | 4,924 | 6,197 | 6,863 | 29,800 |
| URENCO and others | 32,234 | 35,183 | 38,420 | 44,023 | 69,812 |

Source: Department of Finance, Danang City.

government closer to the people and creating space for flexibilities in government programs, budgeting and financing. HCMC is unique in that it has the recycling fund in addition to expenditures for solid waste management. HCMC is the largest generator of municipal solid waste among the cities and provinces, which requires HCMC to expand activities beyond a mere collection and landfill of solid waste, and to promote recycling activities.

Binh Duong Province allocates the highest portion for environmental protection among the three localities. As shown in Table 5.5, from 2006 to present, the shares of environmental expenditures in the total local budget expenditures have been maintained at around 1.5 percent. This accrues to the fiscal decentralization of environmental expenditures to the district level. This is evident from the strong commitment of the provincial government on environmental protection while promoting industrial development at the same time. Binh Duong is the only one among the three cities/provinces in this study which had completed an environmental protection master plan until 2015.

Danang city has been planning to increase environmental expenses in the city government budget in accordance with the "Environmental City" plan initiated in 2008. The environmental expense is expected to reach 1.5 percent in 2015. Most of the environmental expenses went to URENCO for activities related to municipal solid waste management. Besides, environmental expenses by DONRE have been significantly increasing in 2011 to support institutional development in environment. Among those three localities, Danang has not yet set up any environmental fund. Thus, in contrast to Binh Duong where the revenue of environmental protection fees is put into its environmental fund, Danang has allocated this revenue to the budget of the Drainage and Wastewater Treatment Company managed by DONRE.

## Conclusions

During the period 2005–10, environmental protection has been given increased attention from the government at all levels in Vietnam. It is fair to say that this period was a very successful one in developing and improving legislative documents and policies related to environmental protection of the country.

Together with strengthening command and control tools, various types of economic tools have been deployed into environmental management such as environmental protection fees on wastewater, financing to environmental protection through the operation of environmental protection funds at both national and local levels.

At the local level, many provinces and cities have issued regulations on environmental protection for their localities, developed annually and/or five year environmental protection plans, while some have prepared provincial environmental protection strategies toward 2020 and a vision for 2030.

Also in this period, environmental financing and the environmental fiscal mechanism in Vietnam have seen a largely positive change. Before 2005, the government budget expenditures on environment were only 150–200 billion VND each year. Up to now, the government expenses on environmental protection have been rapidly increasing. In 2010, for example, expenses on the environment in the government budget reached about 6,300 billion VND. Along with the change at central government level, the local government expenses at provinces/cities have strongly increased. The environmental management authorities at local level have also taken many initiatives in order to use environmental financing sources more efficiently.

The above mentioned issues provide evidence for the strong commitment of governments at all levels toward environmental policy integration which are also happening in Vietnam to keep up with the sustainable economic development of the country.

While many success stories are recorded in the five year plan 2006–10 in environmental protection and sustainable development, a number of challenges remain yet to be solved. First, despite the priority on the environment in the budget, the total amount is still very limited in comparison with the needs. Environmental expenditures have not been used efficiently enough. Second, environmental fees/charges have been applied for wastewater and solid waste, but existing fees/charges are not efficient and effective yet. Finally, although the environmental tax law came into effect in January 2012, it is still too early to make an assessment of its contribution to environmental pollution control and support of environment expenses allocation.

## Note

1 This implies that the program is closely related to environmental protection/improvement and is funded by government expenditure for development investment.

## Bibliography

Binh Duong Province's People Committee (2011) *Plan on Environmental Protection in Binh Duong 2011–2015*, 2/2011.
Clausen, A., Hoang, H.V. and Pedrono, M. (2011) "An evaluation of the environmental impact assessment system in Vietnam: The gap between theory and practice," *Environmental Impact Assessment Review* 31: 136–43.
Department of Finance, Binh Duong Province (2008, 2009, 2010) *Provincial Government Budget Reports*.

Department of Finance, Danang City (2007, 2008, 2009, 2010, 2011 (*planned*)) *City Government Budget Reports.*

Department of Finance, Ho Chi Minh City (2008 to 2010) *Provincial Budget Reports.*

Department of Natural Resource and Environment, Binh Duong Province (2007, 2008, 2009, 2010) *Annual Budget Expenditure Reports.*

Department of Natural Resource and Environment, Binh Duong Province (2010) *Report on Implementation of Environmental Protection Fees on Wastewater in 2010.*

Department of Natural Resource and Environment, Danang City (2010) *Report on Implementation of Environmental Protection Fees on Wastewater in 2010.*

Ho Chi Minh City Environmental Protection Agency (2010) *Report on Evaluation of Five Year Implementation of Environmental Protection Fees on Wastewater.*

Khoa, L.V. and Ho, P. (2006) "Greening through industrial relocation in Vietnam: The case of Ho Chi Minh City," in Ho, P. (ed.), *Greening Industries in Newly Industrializing Economies: Asian-Style Leapfrogging*, London: Kegan Paul Ltd., 107–35.

Luken, R.A. and Rompaey, F.V. (2007) *Environment and Industry in Developing Countries: Assessing the Adoption of Environmentally Sound Technology*, Cheltenham: Edward Elgar.

Ministry of Finance (2002, 2005, 2006, 2008, 2009, 2010) *State Budget Report.*

Ministry of Natural Resources and Environment (2005, 2006, 2008, 2009) *State of Environment Report.*

# Part II

# Environmental fiscal reform in Europe

# 6 Environmental tax reform

## Experience in Europe and implications for East Asia

*Paul Ekins*

## Environmental tax reform (ETR) in Europe

### *ETR and human well-being*

As noted in the introduction, the European Environment Agency has defined ETR as 'a reform of the national tax system where there is a shift of the burden of taxes from conventional taxes such as labour to environmentally damaging activities, such as resource use or pollution' (EEA 2005, p. 84). ETR is therefore a particular kind of policy instrument, which seeks to apply revenue-raising economic instruments (which may be taxes or auctioned permits in an emissions trading scheme) to resource use and pollution, in order to increase the efficiency of resource use (resource productivity) and improve the environment, and reduce other taxes such that the policy is revenue neutral overall. ETR in this interpretation is therefore a tax *shift*, rather than a tax increase, whereby taxation is shifted from 'goods' such as labour (e.g. income taxes, social security contributions) or capital (e.g. corporation taxes) to 'bads' (pollution, resource depletion). As will be seen, ETR was implemented on a relatively small scale in a number of North European countries in the 1990s and early 2000s, with broadly positive results.

ETR is hypothesised to increase human well-being through the economic and environmental pathways illustrated in Figure 6.1.

This suggests that environmentally ETR will reduce pollution and resource use; economically it will increase output and employment directly, and stimulate green innovation that will do so further indirectly, as well as bringing about further environmental improvement. In themselves, both these economic and environmental impacts will increase human well-being; but the economic impact will be decoupled from environmental damage, so it is more likely to have a positive net impact on human well-being than much current economic growth.

There are good theoretical reasons for supposing that ETR will have the effects shown in Figure 6.1. These have been briefly reviewed in the Introduction to this book, as well as being extensively covered in the environmental economics literature (see Ekins and Barker 2001 for a survey). Very briefly, because many environmental impacts are not priced, they do not enter into market decisions and they are therefore greater, and have a greater negative impact on

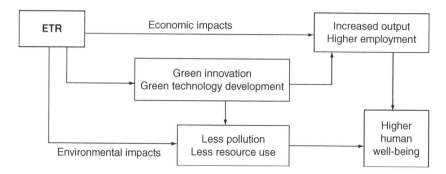

*Figure 6.1* The potential contribution of environmental tax reform to human well-being (source: Ekins 2011, Figure 1.6, p. 15).

human welfare, than is desirable. Environmental taxes give a price (explicitly or implicitly), or a cost, to these impacts, therefore encouraging their reduction, which increases human welfare. This is the rationale for supposing that ETR will lead to increased human welfare from environmental benefits.

The environmental taxes also raise revenues which, for a given level of total tax revenue, allow other taxes to be reduced. Taxes on labour and profits are known to have disincentive effects and cause economic output to be lower than it would otherwise be. Reducing such taxes can therefore, other things being equal, increase output. However, increasing energy taxes, other things being equal, is likely to reduce output. The balance of these two effects (and a key strand of the academic literature explores how different kinds of taxes interact) will determine whether the tax shift will reduce or increase output overall.

A further effect on output may come from the impact of the tax shift on employment. If employers' labour taxes have been reduced as part of the tax shift (and this has often been the case where such shifts have been implemented), this may increase employers' labour demand. If there is unemployment in the economy, this may lead to more people being employed who were formerly unemployed. This would result in an increase in output.

Finally, the shift in the relative prices of carbon/energy and labour or profits will result in a shift in the relative benefits of labour-saving carbon/energy-saving technical change. Because carbon/energy is made relatively more expensive, more technological and innovation effort is likely to be directed to using less of it, thereby raising its productivity. If this process reveals opportunities for carbon/energy saving that were cost effective even before the carbon/energy tax increase, or have become so (and there is substantial evidence of the existence of such opportunities, even in well run and profitable firms), or if it channels innovation into directions that enable other resource savings and product improvements, then the tax shift could also lead to increases in output and employment through this route (the middle pathway in Figure 6.1.

Recycling the revenues from the environmental taxes in an ETR may take effect through reductions in:

- direct taxes (income tax, corporation tax);
- social security contributions

    - paid by employers
    - paid by employees;

- other measures

    - support schemes for investment expenditure (and depreciation); and
    - benefits or other compensatory measures.

An ETR can, in principle, provide complete tax exemptions for economic sectors or reduced tax rates for different energy fuels and economic sectors in combination with some form of negotiated agreements with targets to improve energy efficiency or carbon emissions. Tax ceilings may also be established to limit the total tax burden faced by individual companies. However, such special measures may reduce the economic efficiency of the ETR overall.

Identifying the existence and size of such effects requires a model of the whole economy under consideration, and detailed modelling of the tax shift and its effects on different economic sectors. This chapter describes the results of such modelling of the various ETRs that have been implemented in Europe. It should be borne in mind in this context that models of this sort deliver insights and broad indications of outcomes, rather than 'truth', and that to some extent the outcomes are dependent on modelling assumptions made and the model structure that is chosen. Having said that, provided that the models employed are well grounded in both economic theory and actual data about the European economy and energy system, there is no reason for thinking that their results are not robust.

In recent years three linked projects have investigated the economic and environmental implications of ETR, with special reference to the impacts of ETR on competitiveness, and to the possibility and implications of its implementation on a large scale at the levels of both the European Union and the UK.

### ETR and competitiveness

The first project, called COMETR,[1] investigated the competitiveness effects of environmental tax reforms (see www2.dmu.dk/cometr/ and, for a full exposition of the results Andersen and Ekins 2009).

As already noted, such impacts can only be quantitatively investigated through the use of economic models. These are constructed using both theoretical insights about the relationships between different economic variables (for example, it is normally assumed that the quantity of a good demanded is reduced if its price is increased, and vice versa), and through statistical estimation of the parameters of these relationships. There are different kinds of economic models which make different theoretical assumptions and therefore have different structures. That is one reason why different models can give different outcomes in their modelling of economic interventions such as ETR. This is not the place to go into a detailed comparison of different models, the literature concerning which is now voluminous (for example, see Ekins and Barker 2001). This chapter reports the results of the most comprehensive modelling of European ETRs, which was carried out as part of the European research project COMETR. The model used was a macro-econometric European model (including the 25 countries which were members of the EU in 2006, plus Norway and Switzerland) called E3ME (Economy–Energy–Environment Model for Europe), which is described in some detail in Barker, Junankar *et al.* (2009).

In order to model the effects of ETR in COMETR, a number of scenarios were generated by the macro-econometric model E3ME over the period 1995 to 2012 (the projection period therefore includes Phase 2 of the EU ETS), the main two of which are reported here:

- the Reference Case (R) which is a counterfactual projection without the ETR, but including current and expected developments in the EU economy, e.g. the EU ETS;
- the Baseline Case (B) which is an endogenous solution of E3ME over the period 1995–2012. This scenario includes the ETR in each Member State covered by the project, exemptions or special treatment for the industries most affected and the compensating reduction in another tax. This scenario is calibrated closely to the observed outcome through using historical data which include the effects of ETR implementation.

In the COMETR project, the environmental and economic effects of the ETRs that had been implemented in six EU countries (Denmark, Finland, Germany, Netherlands, Sweden, UK) were modelled. The results are shown in four charts in Figure 6.2, each of which shows the percentage difference between the Baseline and Reference scenarios.[2] As would be expected, given that many of the environmental taxes were applied on the use of fossil fuels, all the six European countries that implemented an ETR show a reduction in fuel demand from the ETR (see Figure 6.2a). In most cases the reduction in fuel demand was in the region of 4 per cent, although it was slightly larger in Finland and Sweden than the other regions.

The scenario results also show that there are reductions in GHGs in all six ETR countries from the ETR (see Figure 6.2b). The effects closely follow the results for total fuel consumption, with the largest reductions occurring in countries with the highest tax rates. The largest reduction in emissions occurs in Finland and Sweden. It should be noted that in most cases the fall in emissions is relatively larger than the fall in fuel demand, indicating that the tax policies, by inducing fuel switching as well as a reduction in energy demand, are efficient at reducing emissions.

All six of the ETR countries show a small increase in GDP as a result of the ETR (see Figure 6.2c). Moreover, the non-ETR countries in the European Union experienced showed practically no change (see Figure 6.2d). This suggests that,

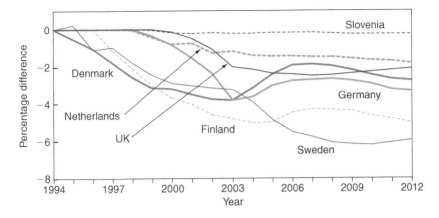

*Figure 6.2a* The effects of ETR on total fuel demand (source: Barker, Ekins *et al.* 2009, Chart 1).

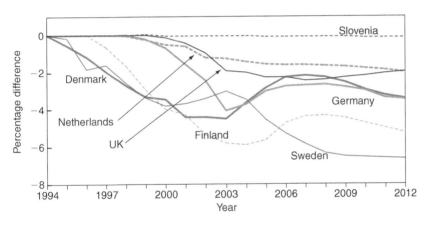

*Figure 6.2b* The effects of ETR on GHG emissions (source: Barker, Ekins *et al.* 2009, Chart 2).

far from damaging the competitiveness of ETR countries compared to the non-ETR countries, the ETR countries benefited economically, as well as environmentally, from the policy.

There are a number of reasons why this might have been the case, given that in itself making energy more expensive would be expected to have a negative effect on energy-using companies, depending on the extent to which they can pass on the energy price increases to their customers, and, at the macro-level, to cause inflation and have a negative effect on economic output.

To begin with, only energy-intensive firms would be significantly affected by the energy price increases, and these account for a relatively small proportion in most economies. Moreover, untraded energy-intensive sectors would be able to

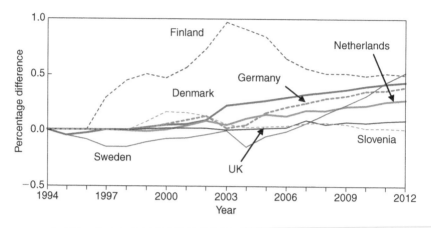

*Figure 6.2c* The effects of ETR on GDP in ETR countries (source: Barker, Ekins *et al.* 2009, Chart 3).

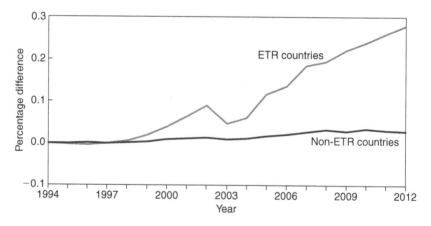

*Figure 6.2d* The effects of ETR on GDP in ETR and non-ETR countries (source: Barker, Ekins *et al.* 2009, Chart 10).

pass on their price increases to consumers, mitigating the competitiveness impacts on the companies concerned.

More importantly, perhaps, firms will seek to reduce their energy use by purchasing energy-efficient intermediate or investment goods from appropriate companies. This could have multiple economic effects. First, it will reduce the energy use of the company making the investment, and this will serve to offset wholly or partly the increased tax expenditures (so that company energy expenditure may actually be *lower* than before the tax increase). Second, it will add to the output of the energy efficiency companies, serving to offset wholly or partly any reduction in output from the increased taxes on energy. Third, the investment will stimulate technical change more generally, especially over the longer term. More energy-efficient equipment is often more productive in other ways as well.

Finally, the revenue recycling mechanism will also affect prices, perhaps directly by reducing the cost of other inputs into production, when this might reduce the prices of goods and services, therefore wholly or partly offsetting the inflationary effect of the tax increase. Another possibility, where the revenue recycling is through a reduction in employers' social security contributions, is that this will increase the demand for labour. Where there is involuntary unemployment, this could increase employment and output over what they would otherwise be. In a situation of full or near full employment, this may act to increase wages, which would then add to the inflationary effect of the tax increase, with further knock-on effects throughout the economy.

All these effects act in different ways on different companies (depending on how their managements respond to the tax increase), different sectors (depending, among other things, on their energy intensities and openness to international trade) and different countries (depending on their overall economic structure). Moreover, there is continuous interaction and feedback at all levels between these effects and all the other influences on economic activity. The effects of ETR on international competitiveness are, therefore, multi-faceted and complex. But the COMETR project found unequivocally that, in respect of the ETRs so far implemented in Europe, these effects have not been significant, but may even have been positive.

### A large-scale ETR for Europe

The second project, called PETRE,[3] explored the subject of 'Resource productivity, environmental tax reform (ETR) and sustainable growth in Europe'. Its final report is at www.petre.org.uk (Ekins 2009a), and a book outcome from the project is Ekins and Speck (2011).

A major objective of the project was to investigate through modelling the implications of a large-scale ETR in Europe. The project used two European macro-econometric models, E3ME and GINFORS, and explored six scenarios:

- Baseline with low energy price (LEP);
- Baseline sensitivity with high energy price (HEP, reference case);
- Scenario 1 (S1(L)): ETR with revenue recycling designed to meet 20 per cent EU 2020 GHG target (scenario compared with LEP baseline);
- Scenario 2 (S1(H)): ETR with revenue recycling designed to meet 20 per cent EU 2020 GHG target (scenario compared with HEP Baseline);
- Scenario 3 (S2(H)): ETR with revenue recycling designed to meet 20 per cent EU 2020 GHG target (scenario compared with HEP Baseline), with a proportion of revenues spent on eco-innovation measures;
- Scenario 4 (S3(H)): ETR with revenue recycling designed to meet 30% 'international cooperation' EU 2020 GHG target (scenario compared with Baseline with HEP).

The taxes in the ETR were as follows:

- A carbon tax rate is introduced to all non-EU ETS sectors equal to the carbon price in the EU ETS that delivers an overall 20 per cent reduction in greenhouse gas emissions (GHG) by 2020 (in the international cooperation scenario (S3(H)) this is extended to a 30 per cent GHG reduction).
- Aviation is included in the EU ETS at the end of Phase 2 in 2012.
- Power generation sector EU ETS permits are 100 per cent auctioned in Phase 3 of the EU ETS (from 2013). (It may be noted that auctioning does not change carbon prices or emissions.)
- All other EU ETS permits are 50 per cent auctioned in 2013 increasing to 100 per cent in 2020.
- Taxes on materials are introduced at 5 per cent of total price in 2010 increasing to 15 per cent by 2020.
- S3(H) carbon tax in non-EU countries is 25 per cent of carbon tax in EU.

The offsetting tax reductions are reductions in income tax rates (for households) and social security contributions (for businesses) in each of the member states, such that there is no direct change in tax revenues. In S2(H) 10 per cent of the environmental tax revenues are recycled through spending on eco-innovation measures.

Table 6.1 shows the results of the modelling. It shows that E3ME projects a small GDP gain from the ETR policy, while GINFORS projects a slightly larger loss. Even the 30 per cent EU $CO_2$ reduction target is reached in GINFORS with the loss of less than 2 per cent of GDP. Both models suggest that ETR will increase employment.

*Table 6.1* Results from the ETR in the PETRE project

| Scenario | $CO_2$ price EURO2008/t | GDP % change from baseline | Employment % change from baseline | Labour productivity % change from baseline |
|---|---|---|---|---|
| S1(L) | | | | |
| E3ME | 142 | 0.6 | 2.2 | −1.6 |
| GINFORS | 120 | −3.0 | 0.0 | −3.0 |
| S1(H) | | | | |
| E3ME | 59 | 0.2 | 1.1 | −0.9 |
| GINFORS | 68 | −0.6 | 0.4 | −1.0 |
| S2(H) | | | | |
| E3ME | 53 | 0.8 | 1.1 | −0.3 |
| GINFORS | 61 | −0.3 | 0.4 | −0.7 |
| S3(H)204 | 204 | 0.5 | 2.7 | −2.1 |
| GINFORS | 184 | −1.9 | 0.8 | −2.6 |

Source: Ekins (2009a, p. 31).

Environmentally the S3(H) scenario has the effect of stabilising global GHG emissions over 2010–2020, as shown in Figure 6.3, while S1(H), the unilateral EU scenario with low energy prices, has a negligible effect on global emissions.

### *A large-scale ETR in a single country (the UK)*

The third recent body of work on ETR to shed light on the effects of this policy instrument was that of the UK Green Fiscal Commission (for its final report, see Ekins 2009b; for a brief description of its operation see Ekins and Shaw, this volume). This was set up in 2007, with its major objective being to investigate the economic, social and environmental implications of a major ETR (GFR), such that the share of environmental taxes in total revenues might rise from 5 per cent to 15–20 per cent in 2020. A further objective was to explore public attitudes to ETR.

As with the PETRE project, the environmental and economic effects of ETR (in this case for the UK) were explored through scenarios, which were defined as follows:

- Three baselines (B1, B2, B3) – medium, low, high world market fossil fuel prices, where the high price, B3, resulted in the same end-user prices of energy as the two ETR scenarios, S1, S2.
- Two ETR scenarios (S1, S2) – increase in transport, household and industrial energy taxes, and taxes on water and materials, reductions in income taxes (households) and social security contributions (business). The taxes in S2 were scaled to give the same end-user energy prices as in S1 (i.e. the taxes were higher because the underlying fossil fuel prices were lower).

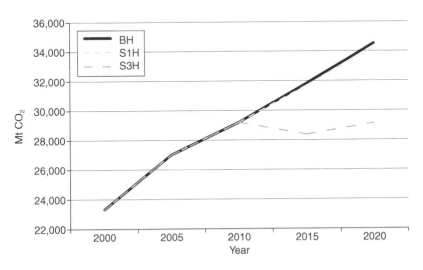

*Figure 6.3* The effect of ETR on global $CO_2$ emissions, GINFORS (source: Ekins 2009a, p. 48).

•   Two 'eco-innovation' scenarios (E1, E2) – spending 10 per cent of green tax revenues on energy-efficient buildings, renewable energy and hybrid vehicles, with the rest of the revenues being recycled as in S1, S2.

Comparing the scenarios S1 and S2 with their respective baselines B1 and B2, and with the results for B3 (which was constructed to have the same end-user fuel prices as S1 and S2), Figures 6.4 and 6.5 first show the scale of the GFR S1 in terms of the shift in the source of tax revenues. In the B1 baseline, tax revenues rise to £875 billion in 2020. Almost 5 per cent of these revenues are accounted for by environmental taxes (£43 billion). Figure 6.4 shows that the composition of the sources of tax revenues does not change substantially over the course of this particular scenario (i.e. between 2006 and 2020).

In S1, however, as a result of the GFR, the share of environmental taxes in tax revenues is higher by design than in B1, at almost 15 per cent (£132 billion). Figure 6.5 shows that the increase in the share of environmental taxes leads to a corresponding decrease (compared to B1) in the share of National Insurance contributions (UK social security contributions) and income tax. This is, of course, to be expected given the specification of the GFR scenarios.

In S2 (not shown here) the share of environmental taxes in all tax revenues in 2020 reaches 17.6 per cent (£155.8 billion). The amount of revenue raised does not differ much between B1 and S1, and B2 and S2, indicating that the tax base arising from GFR over the projection period is stable.

Figure 6.6 shows further results of the modelling, normalised to an index such that the results of the baseline B1 are 100 for both GHG emissions and GDP. It can be seen that the lower fossil fuel prices in B2 result, not surprisingly, in higher GHG emissions and higher GDP. The high world market fossil fuel prices in B3 result in lower GHG emissions, but at a cost of about 6 per cent of GDP.

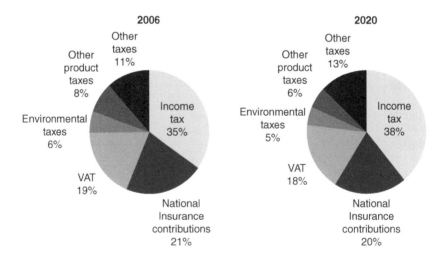

*Figure 6.4* Composition of tax revenues in B1 (source: Ekins 2009b, p. 57).

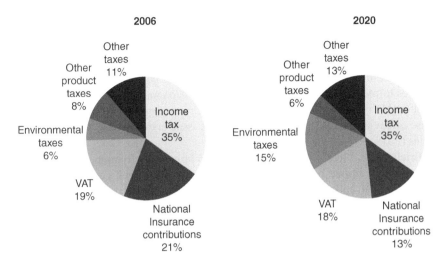

*Figure 6.5* Composition of tax revenues in S1 (source: Ekins 2009b, p. 57).

The ETR scenarios (S1, S2), on the other hand, result in substantial reductions in GHG emissions, but very little change to GDP. The reason for the large difference in the results of B3 and S1 and S2, which have the same high end-user energy prices, is that with B3 the payments for the high prices go outside the UK to energy exporting countries and companies, whereas in the ETR scenarios they are used to reduce other taxes. The 'eco-innovation' scenarios produce useful further reductions in GHG emissions and slightly higher GDP because of the investment effect of the extra innovation spending.

Such results suggest that ETR is a very cost-effective way of reducing GHG emissions and stimulating new eco-industries which could contribute to future

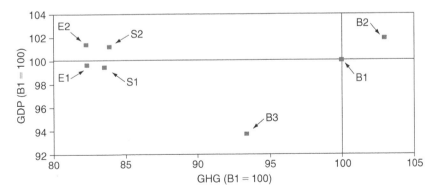

*Figure 6.6* Results for GDP and greenhouse gas emissions (GHG) for different ETR scenarios (source: Ekins 2009b, p. 66).

competitiveness. It would also result in a different trajectory for economic development. It would rule out a resource-intensive growth path, and this would constrain growth unless it led to innovation in low-resource and resource-saving technologies. ETR would stimulate such innovation, but the implementation of complementary policies would probably be desirable to enhance its effect.

The question then arises as to whether ETR and complementary supporting policies to reduce the environmental impact of economic activity would lead to 'sustainable growth' in Europe, and how that rate of growth would compare to those of environment-degrading growth. This issue is explored in some detail in Ekins (2013).

Briefly here, clearly there are doubts as to how long environment-degrading growth can continue before it undermines the environmental conditions necessary for growth and slows down or comes to a halt (that, after all, is the meaning of the word 'unsustainable'). 'Sustainable' growth is likely to be more resource efficient than current and recent growth, and may in time turn out to be slower growth in the short (and perhaps medium) term than 'unsustainable' growth. It would also be likely to have higher employment, implying lower labour productivity and incomes. If the world as a whole moves towards 'sustainable' growth, then the relatively high-growth countries in this world will be those that have developed, and can export, resource-efficient technologies and industries.

ETR emerges from the analysis undertaken as a key policy for fostering sustainable growth. There is certainly no evidence at all that ETR or other policies for environmental sustainability would choke off economic growth altogether. It also seems most unlikely, given the environmental pressures and damages for which there is very clear scientific evidence, that 'unsustainable' growth will last beyond this century, and it could lead to environmental collapse well before 2100. The choice facing economic and other policy makers is therefore clear and from a cost–benefit angle environmental sustainability seems the correct social choice at any but the highest discount rates.

## Implications for East Asia

European experience with ETR cannot be translated wholesale into East Asia, which will need to implement it in a way that is consistent with its fiscal and political context, traditions and conditions. However, the European experience does provide some broad guidance as to how East Asia might proceed with ETR. It starts with the argument that ETR is necessary.

### The imperative of ETR

Most countries in East Asia and elsewhere have now accepted the need to reduce their emissions of GHGs below an unconstrained baseline. If there is to be any chance of keeping average global temperature increases to 2°C, which has now been agreed as the international community's climate change objective, this will require very substantial reductions in emissions of carbon dioxide, the principal greenhouse gas, which are mainly the result of burning fossil fuels. Most of these reductions will need to come from countries with already well-developed economies.

Emissions reductions before 2020 will need mainly to come from investments in renewables and energy efficiency, and reduced demand for energy services, because of the long planning and construction times required for other major low carbon sources of energy supply, nuclear power and carbon capture and storage (CCS). However, all low carbon investments will require clear signals through prices, as well as other policy interventions, that they will be profitable. What is required is a major new policy approach that will send these signals more clearly than at present or in the past.

These signals need to take account of and address some inconvenient facts about energy use which makes reducing carbon emissions difficult. The first is that, other things being equal, energy use increases with income. There is no sustained period of economic growth in any industrial or industrialising society for which this has not been the case. Policy must therefore recognise that, if East Asia is to experience economic growth, as seems very likely, the associated underlying trend of energy use will be increasing. One of the reasons for the close connection between economic growth and energy use is that, consequent to the laws of thermodynamics, energy use is fundamental to all kinds of economic activity (though some activities are obviously more energy-intensive than others). The innovation that drives economic growth is therefore often associated with new products or processes that use energy, what Oreszczyn (2004) has called 'our innate ability to think of new ways to use energy'.

One consequence of this is that improvements in energy efficiency by themselves are most unlikely to reduce the absolute consumption of energy, because although they will reduce the consumption of energy for the delivery of any given service, they will increase the demand for energy services overall and this, combined with the continual creation of new energy service demands through the innovation process, will overwhelm the particular reductions in consumption from the efficiency improvement. Indeed, this must be the lesson of past experience, because industrial societies have seen huge improvements in energy efficiency across practically all uses of energy over the last 50 years, and before, and yet energy consumption has remained obstinately increasing with the size of the economy, as already noted.

The second inconvenient fact is that the only change in economic circumstances that has been shown to reduce energy consumption is an increase in energy prices. If energy prices increase in relative terms, compared with other goods and services, this channels innovation into less energy-intensive routes. If they increase in absolute terms, this reduces the demand for energy services (old and new), and encourages investment in energy conservation and efficiency technologies. If the energy price has been increased through the imposition of a price on carbon (for example, through an emissions trading scheme), then the price increase will also encourage investment in low carbon energy sources.

It was noted above that it is only through increased energy efficiency, increased investment in renewables and reduced demand for energy services that there is any prospect of achieving reductions in carbon emissions in the short term. An increase in energy prices is the only change in economic circumstances that will directly promote all three of these outcomes. It is difficult to avoid the conclusion that such an increase is a necessary (if perhaps not sufficient) condition for carbon reduction on the scale that is now required.

## Ways of increasing energy prices

In the past there have been two sources of energy price increases: from markets, as for example the oil price increases in 1973 and 1979, and more recently in 2007–2008; and from government policy, mainly from taxation such as taxes on industrial or household energy use, or on transport fuels. Such taxes reduce fuel use below what they would otherwise be. In Germany, a 90 per cent increase in diesel prices and 62 per cent increase in petrol prices over 1997–2006, largely driven by increases in taxation, caused the total consumption of the main road fuels to decrease by 13 per cent.

There is a very important economic difference between market-driven and taxation-driven increases in energy prices. In the case of the former the extra revenues accrue to energy companies and energy-producing countries, at the expense of energy-consuming countries. With the latter the government of the energy-consuming country keeps the revenues from the price increase, which, for a given level of government expenditure, allows it to reduce other taxes, with far reduced negative impacts on its economy.

Another important difference is that market-driven increases in, for example, oil prices will stimulate investment into high carbon substitutes for crude oil (for example, oil shale and tar sands), as indeed has happened with the relatively high oil prices over 2006–2008, as well as into low carbon energy sources. Government taxation, in contrast, can target carbon emissions through a carbon tax, which would penalise high carbon oil substitutes, and be far more effective in promoting new investment into low carbon energy sources.

For both these reasons, if it is necessary for the prices of high carbon fuels to increase if carbon emissions are to be reduced, as argued above, governments would be well advised to bring about these increases through taxation, rather than leaving the task to market forces. Recycling the revenues from the taxation by reducing other taxes (or by increasing them by less than would otherwise be the case), is the essence of ETR.

## Political difficulties with environmental taxes

It is highly unfortunate that environmental tax reform (ETR) emerges from the above analysis as a necessary condition for significant carbon reduction, because governments find green taxes politically very problematic. At least four interacting, or mutually reinforcing, factors make this so.

First, people tend to dislike green (and especially energy) taxes more than other taxes, and to regard them as an illegitimate source of general government revenues. There are a number of possible reasons for this: energy taxes affect highly valued forms of consumption (e.g. driving, flying); energy taxes have become regarded as 'stealth' taxes; because energy taxes can be (but do not need to be) regressive, they are regarded as unfair; energy taxes (like other consumption taxes) are not related to ability to pay; some people think that environmental taxes are intended to change behaviour, not raise revenue – revenues deriving from them should therefore be hypothecated back to promote the behaviour change.

Second, people do not trust governments to implement green taxes in a fiscally neutral way. It may also be noted that the desire for revenue neutrality conflicts with a perceived need for hypothecation (which implies an increase in overall taxation). However, both factors – the lack of trust and the demand for hypothecation – tend to limit the politically feasible scope for green taxes.

Third, green taxes on business (like any other taxes on business) can have impacts on competitiveness. As the research reported above has indicated, it is now apparent that, in the case of energy or carbon taxes, the number of sectors that would be significantly affected by even quite a large ETR (which was revenue neutral to business) would be rather small, and the effect on the economy could be positive overall, especially when the taxes incentivise the development of new low carbon industries, as has already happened in Denmark and Germany. But as ever the losers from such a tax are more politically resonant than the potential winners so that competitiveness arguments continue to act as a brake on the implementation of ETR.

Fourth, in some countries energy taxes on households are regarded as regressive and unfair.

As in Europe, all the above factors will need to be addressed to some extent at least if there is to be any chance of introducing a significant ETR, and therefore of reducing carbon substantially, in East Asia. As part of this, it should be stressed that implemented gradually, with appropriate complementary policies, ETR should lead to a number of benefits for the implementing countries apart from its main objective of reduced carbon emissions: new low carbon industries with the possibility of export markets; a more energy-efficient housing stock; the industrial skills required to deploy low carbon and energy efficiency technologies; greater energy security, the fossil fuel importing countries being less vulnerable both to disruptions to supplies of fossil fuels (because of a more diverse energy mix) and to energy price rises in oil and gas markets (because of greater energy efficiency). These benefits make ETR a policy worth fighting for despite its political challenges.

### Implementing ETR

The first priority if ETR is to be implemented is that the argument must be won that it is necessary. This argument has two parts: first that it is imperative to reduce carbon emissions significantly; and second that ETR, entailing steep increases in energy prices, is a necessary policy to achieve such reduction.

Neither part of the argument has yet been convincingly won in most countries. There is not yet a really strong general perception that large-scale reductions in carbon emissions *must* be achieved, and there is even less that ETR is a necessary condition for achieving them.

Once the argument is won in principle that ETR is necessary, a number of conditions will need to be met for it to be able to be implemented in practice. One is that the fiscal neutrality of any ETR will need to be independently monitored to have any credibility. Another condition to be met is that impacts on the competitiveness of vulnerable economic sectors will need to be taken into account and mitigated to the extent possible, without undermining the objectives of the reform. A third is that the ETR will have to be acceptable in terms of its distributional impacts on households. Finally, any ETR would need to be accompanied by further measures, to respond to people's perceptions (as noted above) that environmental tax revenues need to be spent in environmentally enhancing ways to make the tax legitimate (even if this means an increase in government spending overall). These measures would also make it easier for people to adjust to the low-carbon world that the reform is seeking to promote, and therefore both make more acceptable and increase the changes in behaviour which it will bring about.

### ETR as part of a policy package

It has been argued above that ETR is a necessary policy if substantial cuts in carbon emissions are to be achieved. It has not been argued that it is sufficient. Systematic moves towards a low carbon society will require use of the whole policy toolbox – regulation, voluntary agreements, information measures as well as economic instruments, and emissions trading as well as ETR through the use of taxation.

This is not the place to set out in detail the policy package of which ETR could be a necessary part. Here only those complementary policies will be sketched which could play a role in increasing the public acceptability of the ETR itself. First there would be policies to mitigate competitiveness and household distributional impacts, as noted. Second there would be policies to reward perceived good behaviour change. To be revenue neutral these rewards would have to come out of the taxes increased elsewhere. Third, there would be policies to increase the behavioural impact of the ETR in other ways, for example policies which make it easier for people to be aware of their energy use. This is an essential part of the process of moving from a wasteful, cheap energy, high carbon way of life, to much higher energy awareness where low carbon energy is valued much more highly and used with far greater awareness, such that its price may be high but is stable. Fourth, there would be policies which tackled some of the infrastructural barriers to behaviour change (e.g. making it easier to connect distributed energy; public transport improvements). Again, to the extent that this involved expenditure (and it could involve a lot of it), this would either undermine the revenue neutrality or would need a funding source separate from the ETR. Fifth, there could be regulatory policies that made it easier not to use so much energy (e.g. performance criteria on appliances/vehicles/buildings).

Many of these policies have already been introduced in some form in different countries. What is missing is a strong and consistent signal coming from the price of carbon-based energy that it is expensive, and will need to get more so in order to reduce carbon emissions and build the necessary business confidence in the viability of low carbon investments for companies to start mobilising the necessary investment resources at the required scale. ETR could provide such a signal. In addition, carbon prices rising gradually but predictably, and tending to muffle the noise of increasingly volatile energy markets, because market energy prices make up a lower and lower proportion of final energy costs, is a recipe for economic stability and energy security, as well as an ordered transition to a low carbon way of life.

## Notes

1 'COMETR – Competitiveness effects of environmental tax reforms' (2004–2006), FP6 Proposal 501993 funded by DG Research of the European Commission, the support of which is gratefully acknowledged.
2 As the figures show, the analysis also included Slovenia, which introduced a $CO_2$ tax in the 1990s, the first new EU member state to do so. However, it was too small for its modelled impacts to be significant.
3 PETRE was one of four final projects of the Anglo-German Foundation under the collective title 'Creating Sustainable Growth in Europe'.

## References

Andersen, M.S. and Ekins, P. (eds) 2009 *Carbon Taxation: Lessons from Europe*, Oxford University Press, Oxford/New York.

Barker, T., Ekins, P., Junankar, S., Pollitt, H. and Summerton, P. 2009 'The Competitiveness Effects of European Environmental Fiscal Reforms', *European Review of Energy Markets*, Vol. 3, No. 1, pp. 43–75.

Barker, T., Junankar, S., Pollitt, H. and Summerton, P. 2009 'The Effects of Environmental Tax Reform on International Competitiveness in the European Union: Modelling with E3ME', in Andersen, M.S. and Ekins, P. (eds) *Carbon Energy Taxation Lessons from Europe*, Oxford University Press, Oxford, pp. 147–214.

EEA (European Environment Agency) 2005 'Market-based Instruments for Environmental Policy in Europe', EEA Technical Report No. 8/2005, Copenhagen, Denmark.

Ekins, P. 2000 *Economic Growth and Environmental Sustainability: The Prospects for Green Growth*, Routledge, London/New York.

Ekins, P. 2009a 'Resource Productivity, Environmental Tax Reform and Sustainable Growth in Europe', Final Report, Anglo-German Foundation, London, October, http://www.petre.org.uk/pdf/FinRepFin.pdf.

Ekins, P. 2009b 'The Case for Green Fiscal Reform', Final Report of the Green Fiscal Commission, London, October, www.greenfiscalcommission.org.uk/index.php/site/about/final_report/.

Ekins, P. 2011 'Introduction to the Issues and the Book', in Ekins, P. and Speck, S. (eds) *Environmental Tax Reform: A Policy for Green Growth*, Oxford University Press, Oxford, pp. 3–26.

Ekins, P. 2013 'Post-Durban Prospects for Low-Carbon Green Growth' in Chung, S.-Y. (ed.) *Post-2020 Climate Change Regime Formation*, Routledge, London/New York, pp. 68–91.

Ekins, P. and Barker, T. 2001 'Carbon Taxes and Carbon Emissions Trading', *Journal of Economic Surveys*, Vol. 15, No. 3, pp. 325–376.

Ekins, P. and Speck, S. (eds) 2011 *Environmental Tax Reform: A Policy for Green Growth*, Oxford University Press, Oxford.

IPCC (Intergovernmental Panel on Climate Change) 2007 *Fourth Assessment Report*, IPCC, Geneva.

Oreszczyn, T. 2004 'Our Innate Ability to Think of New Ways to Use Energy', *Energy Policy*, Vol. 15, No. 6, pp. 1011–1014.

# 7 Environmental tax reform in the context of recovery from financial crisis

*Stefan Speck*[1]

## Introduction

The concept of environmental tax reform (ETR) – also termed 'green tax reform' – has become well known in European environmental policy. Various ETRs were implemented in the 1990s in Europe and evaluations indicate that they were successful in achieving policy objectives. Nevertheless, such reforms have been fairly limited in number and scope – perhaps because they have been perceived narrowly as an environmental policy tool, based on shifting the burden of taxes between activities and therefore not aimed at generating new revenues.

These limitations were already addressed when international organisations, such as the Organisation for Economic Co-operation and Development and the World Bank, began to promote environmental fiscal reform (EFR) in the early 2000s. Their approach was to promote economic instruments, such as environmental taxes, as a means to generate revenues and further social and environmental goals, for example by boosting spending on health and environmental investments.

Following the economic and financial crises of recent years, politicians are today looking for policy responses that can contribute to the fiscal consolidation process as well as addressing the considerable environmental and climate challenges facing society. Interest in environmental policies, in particular environmental taxes and trading schemes, is arguably increasing because of their potential to perform this dual role. Recent research shows that environmental taxes are essential for the transition to a sustainable or green economy.

This chapter discusses the concept of environmental tax reform and its evolution over time. It highlights how European governments are using environmental taxes to reduce budget deficits and thereby promote the integration of economic and environmental considerations via a sustainable or green fiscal consolidation process.

## ETRs of the 1990s and early 2000s

During the 1990s several European countries implemented environmental tax reforms by shifting the burden of taxation from conventional taxes, for example on labour, to environmentally damaging activities, such as resource use or pollution. This policy approach was promoted in the European Commission's 'White Paper on Growth, Competitiveness and Employment' (EC, 1993), which concluded that the European economy faces a twin challenge wherein 'the "underuse" of labour is combined with an "overuse" of environmental and natural resources'. It further states that 'because the market prices do not incorporate sufficiently the limited availability of those natural resources and the environmental scarcities related to their consumption, their overuse has become systematic'.

At that time, one of the cornerstones of the rationale underpinning environmental tax reform – at least in the literature – was the concept of revenue neutrality, meaning that the overall tax burden on the economy should not be altered.

The then economic situation of European countries may be characterised as a mixture of poor economic performance, a high personal income tax burden and high unemployment; for example Germany faced high unemployment rates and citizens in Sweden and Denmark were burdened with high income tax rates. The ETRs implemented in the countries are reflecting these differences in the economic and fiscal position of European countries (see Speck and Jilkova, 2009).

There is some evidence that the ETRs undertaken in the 1990s and early 2000s were successful in achieving their underlying political objectives, in particular delivering environmental improvements (such as reducing energy consumption and $CO_2$ emissions) while also having negligible or slightly positive effects on economic output and employment (Barker *et al.*, 2009). Clearly, however, *ex-post* analysis of policy effectiveness faces significant challenges in terms of disentangling the effects of a specific policy measure from the impacts of other policies and factors. This is particularly true when seeking to distinguish the effects of an increase in energy and $CO_2$ tax rates from the effects of changing world market prices for energy products.

The tax shifting programmes implemented in Europe during this period shared a broadly similar structure – generating additional tax revenues by extending the tax burden levied on energy products. This was done either by increasing existing excise or energy taxes, introducing new environmental taxes such as $CO_2$ or $SO_2$ taxes, or extending the tax base to include energy products not previously subject to energy taxes.

There were likewise some similarities in national practice in 'recycling' the revenues accrued by reducing other taxes. All countries used a large part of the revenues to reduce taxation of labour, either addressing income tax rates (e.g. Sweden during the 1990s) and/or social security contributions (SSCs) paid by the employer or employee (e.g. Germany and the UK). Furthermore, the countries undertaking ETR also used limited amounts of the revenues generated to support energy efficiency investments.

The impact and duration of these recycling measures varied between countries, with implications for public perceptions of ETR as a whole. In Sweden the reduction of income tax rates had a long-lasting effect. It should be recognised, however, that Sweden implemented a major fiscal reform programme in the

1990s with the objective of reducing income tax rates. The revenues generated from environmental taxes only partially offset the loss of revenues resulting from reduced income tax rates, meaning that the reforms did not adhere to the revenue neutrality principle. This was in strict contrast to the policy approach adopted in Germany and the UK.

Assessing the effectiveness of recycling revenues via the reduction of SSCs is not as straightforward. In Germany and the UK, for example, the reduction of the SSC rates was soon reversed in response to the changing macroeconomic situation (Speck, 2007). In Germany, pension fund contributions paid by employers and employees have been reduced (Table 7.1) as part of the ETR starting in April 1999.

Introducing the ETR did mean that subsequent increases in SSCs which commenced in 2003 as a consequence of the overall economic situation started from a lower base than might otherwise have been the case. Nevertheless, where reforms result in only a temporary cut in SSCs (as occurred in Germany and the UK) it may be harder to convince the public of the benefits of an ETR. Vested interests, for example energy-intensive industries, which regularly oppose energy and $CO_2$ taxes because of concerns about losing competitiveness, can point to this temporary character as an argument against the reform. It is far from straightforward to convince citizens that they would be facing higher SSCs without the ETR, whereas higher energy prices resulting from the ETR may be much more obvious.

*Table 7.1* Development of the rate of contribution to social insurance in Germany (expressed as a percentage of total earnings)

|  | Pension fund | Health insurance | Unemployment insurance | Nursing/care insurance | Total |
|---|---|---|---|---|---|
| 1995 | 18.6 | 13.2 | 6.5 | 1 | 39.3 |
| 1999 – until March | 20.3 | 13.6 | 6.5 | 1.7 | 42.1 |
| 1999 – from April | 19.5 | 13.6 | 6.5 | 1.7 | 41.3 |
| 2000 | 19.3 | 13.6 | 6.5 | 1.7 | 41.1 |
| 2001 | 19.1 | 13.6 | 6.5 | 1.7 | 40.9 |
| 2002 | 19.1 | 14.0 | 6.5 | 1.7 | 41.3 |
| 2003 | 19.5 | 14.3 | 6.5 | 1.7 | 42.0 |
| 2005 | 19.5 | 14.6 | 6.5 | 1.7 | 42.3 |
| 2007 | 19.9 | 14.8 | 4.2 | 1.7 | 40.6 |
| 2010 | 19.9 | 14.9 | 2.8 | 1.95 | 39.6 |
| 2011 | 19.9 | 15.5 | 3 | 1.95 | 40.4 |
| 2012 | 19.6 | 15.5 | 3 | 1.95 | 40.1 |

Source: Speck (2007) and UDE (2013).

Note
The payments of social security contributions (SSC) are split equally between employer and employee except that since 2009 households have had to pay an additional 0.9% for health insurance and since 2005 childless households have to pay an additional 0.25% for the nursing/care insurance, which is not included in the table.

One of the obstacles to communicating the benefits of ETR is that the revenues generated for reducing SSCs are small when compared to the total SSC payments. For example, in Germany total SSCs amounted to €364 billion in 2003, while energy tax revenues totalled about €50 billion, with roughly €19 billion earmarked for recycling measures, i.e. about 5 per cent of total SSC revenues.

In general, experience suggests that earmarking specific ETR revenues to offset other taxes and charges in a revenue neutral way, as was done in Germany and the UK, can undermine/challenge the case for ETR, i.e. the continuing reduction of SSC as the overall economic development can reverse the reduction. The Swedish example differs significantly since the government did not explicitly promote it as a tax-shifting programme.

The ETR discussion has changed in recent years, as international organisations developed the concept of environmental fiscal reform as a policy measure appropriate for both developed and developing countries (World Bank, 2005; OECD, 2005). EFR is a broader concept and summarised by the OECD (2005) as follows:

> 'Environmental fiscal reform' (EFR) refers to a range of taxation and pricing measures which can raise fiscal revenues while furthering environmental goals. This includes taxes on natural resource exploitation or on pollution. EFR can directly address environmental problems that threaten the livelihoods and health of the poor. EFR can also free up economic resources or generate revenues that can help to finance access of the poor to water, sanitation and electricity services.

Defined in this way, fiscal neutrality is not considered a pivotal component of EFR. It is a broader concept than ETR in the sense that it embraces the reform and removal of environmentally harmful subsidies and the use of other pricing measures such as emission trading schemes. The EU emissions trading scheme – an important tool in climate change policy at the EU level – potentially provides an opportunity to generate additional revenues as auctioning of emission allowances will become the rule in the third phase (2013–2020). The extension of the comparatively narrow definition of ETR, in particular as a pure tax-shifting policy, is therefore timely.

## Environmental tax reform – the latest development

Notwithstanding the current political and economic situation, the concept of ETR persists and retains its merits. For example, the European Commission states that

> The tax burden on labour should be substantially reduced in countries where it is comparatively high and hampers job creation. To ensure that reforms are revenue-neutral, taxes such as consumption taxes, recurrent property taxes and environmental taxes could be increased.

(EC, 2012a)

This is reflected in recent policy development in countries such as Denmark and Sweden, which introduced new ETRs during 2009. Outside Europe, countries like Australia and South Africa have also introduced tax-shifting policies (Speck, 2010).

---

**Box 7.1 Danish tax reform – including an ETR component**

A major tax reform is being phased in from 2010 to 2019 with the aim of reducing the fiscal burden on personal income in order to stimulate labour supply in the long term. The reform is designed so as to guarantee overall revenue-neutrality, while the timing for introducing the different measures takes into account cyclical conditions...

In line with the other energy and climate policy objectives of the government, financing of the reform is partly provided by higher energy, transport and environmental taxes, and also by increases of excise rates on health-related goods, such as tobacco and candy. In particular, energy taxes on business and households – except for petrol and diesel – are increased by 15%. To minimise the negative effect on household disposable income, a lump-sum transfer ('green check') will be granted to adults and children as well (up to two per household). The 'green check' is nominally fixed (DKK1 300 for adults and DKK300 for children) and is rapidly phased out for income above DKK360 000 (€48 300).

Source: EC (2010)

---

When assessing the development of EU-27 tax revenues relative to GDP it can be concluded that since 2005 this ratio first increased slightly and dropped by one percentage point between 2007 and 2010. In contrast, environmental taxes increased slightly as a percentage of total tax revenues between 2008 and 2010, reversing the downward trend since 2004 (see Table 7.2). The ratio between environmental taxes and GDP increased between 2008 and 2009 and remained constant between 2009 and 2010. Nevertheless, in 2010 environmental tax revenues were still lower relative to GDP and total tax revenues than in the late 1990s.

The shift to auctioning of emission allowances under the EU Emission Trading System from 2013 will boost the revenues from economic instruments related to environmental policies. This is likely to be reflected in an increase in the ratio of environmental taxes to GDP and to total tax revenues. Some countries have already announced, however, that these revenues will be hypothecated for climate and energy funds (Germany) or for environmental funds (Czech Republic), which means that the revenues will not be available to offset other taxes.

During the period 2004–2010, environmental tax revenues increased considerably in the new EU Member States, achieving an annualised nominal growth rate of more than 8 per cent in Bulgaria, Estonia and Romania (Table 7.3). In contrast the environmental tax revenues remained constant or decreased in nominal terms in countries like Denmark, Germany, Ireland and Portugal, as well as in Cyprus.

*Table 7.2* Development of EU-27 tax revenues, 1995–2010

| | 1995 | 1998 | 1999 | 2000 | 2001 | 2002 | 2003 | 2004 | 2005 | 2006 | 2007 | 2008 | 2009 | 2010 |
|---|---|---|---|---|---|---|---|---|---|---|---|---|---|---|
| Total taxes (including SSC) as % of GDP – arithmetic average | 36.6 | 37.1 | 37.2 | 37.0 | 36.4 | 36.1 | 36.2 | 36.2 | 36.6 | 36.7 | 37.2 | 36.7 | 35.8 | 35.6 |
| Environmental taxes as % of total taxation – arithmetic average | 7.4 | 8.1 | 8.1 | 7.6 | 7.6 | 7.6 | 7.8 | 8.0 | 7.7 | 7.3 | 7.1 | 7.0 | 7.3 | 7.4 |
| Environmental taxes as % of GDP – arithmetic average | 2.7 | 2.9 | 2.9 | 2.8 | 2.7 | 2.7 | 2.8 | 2.8 | 2.8 | 2.7 | 2.6 | 2.5 | 2.6 | 2.6 |

Source: Eurostat (2012).

Table 7.3 Overview of changes in total tax and environmental revenues and GDP in nominal terms (%)

| | Annualised change environmental tax revenue 2004-2010 (nominal prices) | Annualised change environmental tax revenue 2004-2010 (in constant 2005 prices) | Annualised change total tax revenue (incl. SSC) 2004-2010 (nominal prices) | Annualised change total tax revenue (incl. SSC) 2004-2010 (in constant 2005 prices) | Annualised change GDP 2004-2010 (nominal prices) | Percentage change environmental tax revenue 2008-2010 (nominal prices) | Percentage change total tax revenue 2008-2010 (nominal prices) |
|---|---|---|---|---|---|---|---|
| Austria | 1.4 | -0.5 | 3.0 | 1.1 | 3.4 | 1.2 | -0.4 |
| Belgium | 1.8 | -0.5 | 3.0 | 0.7 | 3.3 | 7.5 | 1.7 |
| Bulgaria | 8.3 | 1.8 | 7.0 | 0.6 | 10.1 | -13.3 | -13.7 |
| Cyprus | 0.0 | -2.2 | 6.9 | 4.5 | 5.5 | -8.4 | -6.6 |
| Czech Republic | 3.6 | 1.1 | 3.3 | 0.8 | 4.3 | -1.9 | -3.6 |
| Denmark | 0.0 | -2.0 | 2.5 | 0.5 | 3.0 | -4.7 | -0.4 |
| Estonia | 13.3 | 8.1 | 8.7 | 3.8 | 6.7 | 14.5 | -5.3 |
| Finland | 0.6 | -1.2 | 2.3 | 0.5 | 2.9 | 0.7 | -4.7 |
| France | 0.8 | -0.9 | 2.3 | 0.6 | 2.6 | 0.0 | -1.6 |
| Germany | -0.8 | -2.4 | 1.9 | 0.3 | 2.0 | 0.1 | -1.9 |
| Greece | 5.0 | 1.6 | 3.3 | 0.0 | 3.5 | 17.1 | -6.0 |
| Hungary | 2.5 | -2.4 | 4.4 | -0.6 | 4.4 | -3.0 | -5.7 |
| Ireland | -0.1 | -1.3 | -0.4 | -1.7 | 0.6 | -16.8 | -17.4 |
| Italy | 0.6 | -1.5 | 2.6 | 0.5 | 1.8 | 2.7 | -2.1 |
| Latvia | 7.9 | 1.1 | 8.5 | 1.7 | 9.3 | -4.4 | -25.5 |
| Lithuania | 1.0 | -3.6 | 6.5 | 1.6 | 7.1 | 0.7 | -23.4 |
| Luxembourg | 2.1 | -0.5 | 6.5 | 3.7 | 6.6 | -2.0 | 6.7 |
| Malta | 5.8 | 3.3 | 5.5 | 3.1 | 5.2 | -3.5 | 4.3 |
| Netherlands | 3.5 | 2.0 | 3.6 | 2.1 | 3.1 | 1.5 | -2.0 |
| Poland | 7.3 | 4.4 | 7.5 | 4.6 | 7.3 | 10.9 | 2.8 |
| Portugal | -0.5 | -2.3 | 3.0 | 1.2 | 2.5 | -3.4 | -3.5 |
| Romania | 9.7 | 2.9 | 12.2 | 5.2 | 12.2 | 1.8 | -15.2 |
| Slovakia | 6.7 | 4.2 | 9.6 | 7.0 | 11.7 | -3.0 | -1.7 |
| Slovenia | 6.0 | 3.1 | 4.4 | 1.5 | 4.5 | 14.0 | -3.0 |
| Spain | 0.0 | -2.5 | 2.3 | -0.3 | 3.8 | -3.3 | -6.6 |
| Sweden | 3.7 | 1.8 | 2.9 | 1.0 | 3.7 | 6.9 | 1.8 |
| United Kingdom | 3.3 | 0.6 | 3.5 | 0.8 | 3.3 | 10.0 | -4.7 |

Source: author's calculations based on Eurostat (2012) and the harmonised index of consumer prices (HCIP) (2005 = 100) published by Eurostat is used for calculating the real annualised changes.

Note
The calculation is based on revenues in national currencies.

The situation is mixed when assessing the annualised changes of environmental tax and total tax revenues in constant prices (in 2005 prices) as in 14 out of the 27 EU Member States the environmental tax revenues were reduced during the period between 2004 and 2010 as compared to only three EU Member States in which the total tax revenues including SSC dropped.

Revenues from environmental taxes can be described as relatively stable during the financial crisis (i.e. between 2008 and 2010) in the sense that in 22 of the 27 EU Member States they did not fall as much as total tax revenues. The large drop in Denmark was due to a large fall in revenues from motor vehicle registration tax. The development in Greece deserves special attention as environmental tax revenues increased by more than 17 per cent during this period as a consequence of the sharp increase in the tax rates of transport fuels (petrol and diesel) as shown in Table 7.4 (below).

It should be noted that the European Commission is today one of the most forthright proponents of ETR, as the following statement makes clear (EC, 2011):

> At any given level of the overall tax burden, tax systems should be reviewed to make them more employment-, environment- and growth-friendly, for example via 'green tax reforms' which consist of increasing environmental taxes while reducing other more distortionary taxes.

This testimony is of great significance as it affirms that environmental taxes are less distortionary than other taxes.

## Future prospects for ETR and EFR

Today countries in Europe and globally face a combination of complex economic, social and environmental challenges (EEA, 2010). The Stern Review (2007) clearly set out the urgency of strong and early action on climate change, as well as the potential benefits. Carbon pricing either through a carbon tax or emission trading is widely accepted as a cornerstone in climate change policy. Furthermore, economic instruments are seen as the enabling conditions for the transformation towards a green economy (UNEP, 2011; OECD, 2011; Ekins and Speck, 2011). The urgency of this transition is becoming more apparent because of the existence of multiple and often interconnected crises in areas such as finance, energy, climate change and demography.

Countries evidently face different conditions compared to the situation a decade or two ago, when several European countries introduced ETRs. Many European countries have to address substantial fiscal deficits as a consequence of the financial crisis. Concerns about how countries will deal with the economic and fiscal implications of ageing populations are also growing (EC, 2009).

Efforts to address these concerns need to take into account current discussions about transitioning towards a green economy. The green economy/green growth debate highlights the need to integrate environmental realities into economic policy, as economic growth is still seen as the means to overcome today's challenges. Equally, this debate must take the financial challenges seriously, including the need to achieve fiscal sustainability in national budgets. In this context, it is important to distinguish between the unbalanced budget, i.e. the government budget deficit, and the debt of a country as measured as the ratio of the gross government debt to GDP.[2] Many countries have implemented concrete steps to overcome their fiscal deficits, reflecting the widespread recognition that deficits have to be reduced by addressing both the expenditure revenue sides of the public accounts (see e.g. IMF, 2010a). Spending cuts implemented in austerity packages throughout Europe are part of the debt consolidation strategies. But these spending cuts are and should be accompanied by additional tax revenues. For example, Baldacci *et al*. (2010), referring to an IMF (2010b) report state:

> Unlike previous research on fiscal consolidation, our findings show that raising tax revenue is key to successful debt reduction in countries with large fiscal adjustment needs. This reflects the need to maintain a balance between expenditure savings and revenue-raising measures...
>
> Measures to increase taxation should, however, be designed in a way that does not harm efficiency and minimizes distortion, particularly where taxes as a percentage of GDP are already high. Simplifying the tax system by reducing excessive tax rates and broadening the tax base could help enhance revenue collection while shifting the burden of taxes away from productive inputs. For example, financial sector and carbon taxation may help the budget while at the same time addressing efficiency concerns.

The question of how to split the contribution of spending cuts and increased tax revenues in a debt consolidation strategy is still debated. As noted in *The Economist* (2010), a report 'published by Policy Exchange, a right-of-centre British think-tank, concluded that successful debt consolidations had put 80% of the emphasis on spending cuts'.

To the extent that tax increases are employed, it is clearly important that any economic instruments introduced lead to a more efficient fiscal system. Stiglitz (2010) provides some general principles for deficit reduction, notably: 'It is better to tax bad things (like pollution) than good things (like work)' and 'Economic stability requires environmental stability. The polluter pays principle – making polluters pay for the costs they impose on others – is good both for efficiency and for equity.'

Stiglitz also stresses that taxes discouraging activities that generate negative externalities can increase economic efficiency and that

> the most important are those associated with carbon emissions, with their impact on global warming and climate change. It matters less whether those generating the pollution pay a carbon tax or buy emission permits that are auctioned; either can generate large amounts of money and simultaneously improve economic performance.
>
> (Stiglitz, 2010)

Recent years have seen a growing interest in assessing the potential for taxes to boost economic growth. This interest can also be attributed to the deterioration of public finance as well as the emergence of the concept of green growth. Research has shown that consumption taxes that include environmental taxes and value added taxes are 'the least detrimental to growth' (EC, 2010). As noted in that report, Heady *et al.* (2009) conclude 'that a revenue-neutral shift of taxes from income taxes to consumption and property taxes of 1% of GDP would increase GDP per capita in the range of ¼ to 1 percentage point in the long run'.

Some EU Member States that are heavily affected by the financial crisis, such as Ireland and Greece, have increased the tax burden on energy products. Table 7.4 shows the relative increases in tax rates levied on transport fuels (petrol and diesel) between 2008 and 2011. More recently Italy increased petrol and diesel tax rates as part of the austerity package approved by the Italian parliament in mid-July 2011. The revenues accrued through these increases are being used for the fiscal consolidation.

When examining the development of taxes levied on mineral oil products (petrol and diesel) between January 2008 and July 2011, it must be stated that all 27 EU Member States with the exception of Germany and Portugal increased one or more of their tax rates levied on transport fuels. In addition to the increase in energy taxes, several countries raised their value-added tax (VAT) rates, which also affect energy prices.

These developments are interesting when viewed alongside the difficulties that governments have faced in applying energy and $CO_2$ taxes more widely due to concerns about distributional and competitiveness impacts. Both issues have been studied in detail and it has been demonstrated that such concerns can be addressed and at least partly offset when energy and $CO_2$ taxes are developed as part of an ETR policy package.[3] It is indisputable that energy and $CO_2$ taxes have regressive effects but it is important to keep in mind that an increase in VAT also has negative distributional implications. Since the beginning of 2008 a total of 17 EU member states increased VAT and the highest increases can be found in Hungary (up 7 p.p. to 27 per cent), Romania (up 5 p.p. to 24 per cent), the UK (up 2.5 p.p. to 20 per cent) and Greece (up 4 p.p. to 23 per cent).

A quite different conclusion must be drawn regarding the revenues raised from transport fuels during the period 2008–2011, as the revenues increased in some EU Member States considerably, while falling in others. The most interesting countries are Greece, which reports an increase of 44 per cent in total tax revenues from sales of transport fuels. This rise is in some respects unsurprising since Greece also reported the biggest increase in tax rates, although it has also coincided with a period when Greece was confronting huge political and economic crises.

Other EU Member States with large increases in transport fuel tax revenues are Romania (30 per cent), Malta (27 per cent) and Cyprus (24 per cent). On the other hand the transport fuel tax revenues in Portugal, which was also confronting economic crisis dropped by 14 per cent during this period, followed by Latvia (down 9 per cent) and Spain (down 8 per cent).

*Table 7.4* Percentage changes in energy tax rates levied on petrol (unleaded) and diesel (propellant) between January 2008 and December 2011 and changes in nominal revenues of taxes levied on petrol (unleaded) and diesel between 2008 and 2011

| | Tax rates | | Tax revenues transport fuel | | |
| --- | --- | --- | --- | --- | --- |
| | *Petrol* | *Diesel* | *Petrol* | *Diesel* | *Total* |
| Austria | 8 | 13 | n.a. | n.a. | n.a. |
| Belgium | −1 | 35 | −17 | 26 | 13 |
| Bulgaria | 4 | 3 | −17 | 0 | −6 |
| Cyprus | 20 | 35 | 13 | 40 | 24 |
| Czech Republic | 8 | 10 | −7 | 3 | −1 |
| Denmark | 4 | 6 | −8 | −1 | −4 |
| Estonia | 18 | 19 | 3 | 13 | 8 |
| Finland | −1 | 10 | −8 | 10 | −1 |
| France | 1 | 3 | −14 | 7 | 1 |
| Germany | 0 | 0 | −3 | 8 | 2 |
| Greece | 91 | 41 | 60 | 21 | 44 |
| Hungary | 13 | 11 | −6 | 11 | 4 |
| Ireland | 23 | 22 | 1 | 18 | 9 |
| Italy | 9 | 12 | n.a. | n.a. | n.a. |
| Latvia | 38 | 31 | −17 | −5 | −9 |
| Lithuania | 34 | 10 | −17 | 7 | −2 |
| Luxembourg | 0 | 3 | −12 | 10 | 4 |
| Malta | 16 | 15 | 21 | 33 | 27 |
| Netherlands | 3 | 12 | −1 | 6 | 2 |
| Poland | 1 | 8 | n.a. | n.a. | 10 |
| Portugal | 0 | 0 | −16 | −12 | −14 |
| Romania | 40 | 40 | 13 | 39 | 30 |
| Slovakia | 7 | −20 | n.a. | n.a. | n.a. |
| Slovenia | 16 | 16 | 8 | 21 | 16 |
| Spain | 8 | 10 | −11 | −7 | −8 |
| Sweden | 4 | 9 | −10 | 20 | 2 |
| United Kingdom | 15 | 15 | 0 | 17 | 9 |

Source: author's calculation based on EC DG Energy (Oil Bulletin) and DG TAXUD.

Note
The calculation is based on tax data in national currencies; the tax revenue figures for Denmark and Germany cover the period 2008–2010. For Cyprus they cover 2009–2011. Petrol tax revenues for Greece include data for leaded and unleaded petrol.

Studies assessing the long-term sustainability of public budgets are attracting increased attention, in particular as in view of the growing pressure on public spending expected as a consequence of population ageing.[4] One of the recurrent themes in this discussion is the possibility of using carbon pricing to generate a revenue stream that could be used to offset some of the increase in public spending. Modelling by the IMF (2010b) suggests that 'efficient carbon-pricing schemes could raise 0.75 percent of GDP in advanced economies and 1.5 percent of GDP in emerging economies within the next ten years, while targeted transfers could offset any impact on the poor'.

Nobody expects that carbon pricing will raise enough revenues to cover the increase in expenditures resulting from demographic change. It can, however, provide some of the revenues needed in the short and medium term. Moreover, recent research by Vivid Economics (2012) concludes that: 'energy taxes would cause less economic harm per unit of revenue than direct (i.e. income) or indirect taxes, while also producing other benefits'. The significance of this statement must be seen in the context of fiscal consolidation strategies.

The economic and financial situation in several EU Member States necessitates vast efforts to consolidate current budgets and reduce public debt. At the same time policy measures to initiate the transition towards a green and sustainable economy are crucial. Environmental taxes are effective in this situation as they can serve an environmental purpose but simultaneously generate fiscal revenue. As the European Commission (2012b) has noted: 'Fiscal consolidation, which reduces the scope for environmental policy measure on the expenditure side of the budget, strengthens the need to use [environmental] taxes as well as other market-based policy instruments in environmental policy.'

## Summary

Environmental tax reform as a programme for shifting the burden of tax from 'goods' such as human endeavour to 'bads' such as pollution certainly has a role in fiscal reform discussions. The ETRs implemented so far have shown that the underlying rationale as well as the anticipated policy objectives can be achieved, in particular when the ETR is embedded in a larger fiscal reform package as was the case in Sweden and thereby not adhering to the revenue neutrality principle. The Danish government has followed this approach, lowering the fiscal burden on personal income and offsetting the resulting loss of revenues via increased environmental taxes, which also help achieve national energy and climate policy objectives.

Clearly, however, the tax-shifting approach may not be appropriate in countries facing significantly different economic and fiscal situations. Countries like Greece and Ireland, for example, must identify new income streams for state budgets and have decided to raise – dramatically – the taxes levied on energy products as part of their fiscal consolidation process. This implies that additional revenues cannot be used to reduce other taxes according to the principle of revenue neutrality, which is one of the underlying rationales of ETR. Instead, they are channelled to the general budget, following the broader principle of environmental fiscal reform (EFR).

The EFR concept offers a mixture of benefits, most obviously the prospect of helping make public finances sustainable. State spending is very likely to increase over time because of the demographic changes that many European countries face in coming decades. In this context, revenues from environmental taxes can provide a useful means of covering these expenditures. As the European Commission notes, 'the evidence … indicates that environmental taxation deserves careful consideration, particularly in the current context of revenue shortfall' (EC, 2010).

Another key benefit of EFR is its capacity to correct market prices to reflect the full costs of producing and consuming goods and services. As the Stern Review notes, climate change is the greatest and widest ranging market failure ever seen, presenting a unique challenge for economics and necessitating government intervention. Economic theory and political reality demonstrate that environmental taxes are a means of correcting this market failure.

Next, EFR should be understood in the context of the transformation from a 'brown' economy into a 'green' one, which has become an important theme in international governance during recent years. One of the essential tools for starting this process is carbon pricing either in the form of carbon taxes or via emission allowances.

Finally, in view of the manifold benefits that it offers, it is clear that fiscal reforms be undertaken in a manner that engenders public support so far as possible. By avoiding a direct link between the introduction or augmentation of environmental taxes and reductions to other taxes and charges, EFR eliminates the need to defend increases in the latter taxes and charges if they are necessitated by macroeconomic or demographic realities (as was the case in Germany).

## Notes

1 European Environment Agency, Copenhagen, Denmark. Disclaimer: the views expressed in this chapter are those of the author and may not in any circumstances be regarded as stating an official position of the European Environment Agency or its Management Board.
2 See for a more detailed discussion on this topic Ekins and Speck (2011).
3 See, for example, Ekins and Speck (1999, 2008); Barker *et al.* (2009) and Speck and Jilkova (2009) regarding potential loss of competitiveness. See EEA (2011) regarding distributional implications.
4 See for example EC (2009).

## References

Baldacci, E., Gupta, S. and Mulas-Granados, C. (2010) 'Getting Debt under Control', *Finance and Development*, 47 (4): 18–21.
Barker, T., Junankar, S., Pollitt, H. and Summerton, P. (2009) 'The Effects of Environmental Tax Reform on International Competitiveness in the European Union: Modelling with E3ME', in: M.S. Andersen and P. Ekins (eds), *Carbon-Energy Taxation: Lessons from Europe*, Oxford: Oxford University Press, 147–214.
Ekins, P. and Speck, S. (1999) 'Competitiveness and Exemptions from Environmental Taxes in Europe', *Environmental and Resource Economics*, 13(4): 369–396.
Ekins, P. and Speck, S. (2008) 'Environmental Tax Reform in Europe: Energy Tax Rates and Competitiveness', in: L. Kaiser, J. Milne and N. Chalifour (eds), *Critical Issues in Environmental Taxation: Volume V*, Oxford: Oxford University Press, 77–105.
Ekins, P. and Speck, S. (2011) 'The Fiscal Implications of Climate Change and its Policy Response', Paper prepared for the UNEP MCA4 climate initiative, www.mca4climate.info/_assets/files/ClimatePolicy_FiscalSustainability_Final_Report(1).pdf (accessed 27 January 2013).

European Commission (EC) (1993) 'Commission White Paper on Growth, Competitiveness and Employment', COM(93)700 final, Brussels.

European Commission (EC) (2009) '2009 Ageing Report: Economic and Budgetary Projections for the EU-27 Member States (2009–2060)', European Economy 2/2009.

European Commission (EC) 2010, 'Monitoring Tax Revenues and Tax Reforms in EU Member States 2010: Tax Policy After the Crisis', Brussels.

European Commission (EC) (2011) 'Annual Growth Survey Annex 2: Macro Economy Report', COM(2011)11 final Annex 2, Brussels.

European Commission (EC) (2012a), 'Annual Growth Survey 2013', Communication from the Commission, COM(2012)750 final, Brussels.

European Commission (EC) (2012b) 'Tax Reforms in EU Member States 2012: Tax Policy Challenges for Economic Growth and Fiscal Sustainability', European Economy 6/2012, DG for Economic and Financial Affairs and DG for Taxation and Customs Union, Brussels.

European Environment Agency (EEA) (2010) *The European Environment State and Outlook 2010 Synthesis*, Copenhagen: EEA.

European Environment Agency (EEA) (2011) 'Environmental Tax Reform in Europe: Implications Income Distribution', EEA Technical report No. 16/2011, Copenhagen.

Eurostat (2012) 'Taxation Trends in the European Union Data for the EU Member States, Iceland and Norway, 2012 Edition', Report prepared by Eurostat and European Commission, Taxation and Customs Union, Luxembourg, Publication Office of the European Union.

Heady, C., Johansson, A., Arnold, J., Brys, B. and Vartia, L. (2009) 'Tax Policy for Economic Recovery and Growth', University of Kent, *School of Economics Discussion Papers* 0925.

International Monetary Fund (IMF) (2010a) 'From Stimulus to Consolidation: Revenue and Expenditure Policies in Advanced and Emerging Economies', Paper prepared by the Fiscal Affairs Department, Washington.

International Monetary Fund (IMF) (2010b) 'Fiscal Exit: From Strategy to Implementation', Fiscal Monitor, Washington.

Organisation for Economic Co-operation and Development (OECD) (2005) 'Environmental Fiscal Reform for Poverty Reduction', DAC Guidelines and Reference Series, DAC Guidelines and Reference Series, Paris.

Organisation for Economic Co-operation and Development (OECD) (2011) *Towards Green Growth*, Paris: OECD.

Speck, S. (2007) 'Differences in ETR between CEEC and Germany/UK', Paper presented at the Mid-Term Conference of the Anglo-German Foundation project 'Resource Productivity, Environmental Tax Reform and Sustainable Growth in Europe', Berlin, 2007.

Speck, S. (2010) 'Options for Promoting Environmental Fiscal Reform in EC Development Cooperation: Country Case Study South Africa', case study of the project 'Fiscal Reform in Development Cooperation' funded by EuropeAid Cooperation Office and EC DG Development, www.foes.de/pdf/2010-09-29%20South%20Africa%20-%20 FINAL.pdf (accessed 27 January 2013).

Speck, S. and Jilkova, J. (2009) 'Design of Environmental Tax Reforms in Europe', in: M.S. Andersen and P. Ekins (eds), *Carbon-Energy Taxation: Lessons from Europe*, Oxford: Oxford University Press, 24–52.

Stern, N.H. (2007) *The Economics of Climate Change: The Stern Review*, Cambridge: Cambridge University Press.

Stiglitz, J.E. (2010) 'Principles and Guidelines for Deficit Reduction', The Roosevelt Institute, Working Paper No. 6, 2 December.

*The Economist* (2010) 'Sharing the Pain', 4 March.

Universitaet Duisburg Essen (UDE) (2013) 'Beitragssatzentwicklung in der Sozialversicherung 1970–2013', www.sozialpolitik-aktuell.de/tl_files/sozialpolitik-aktuell/_Politik-felder/Finanzierung/Datensammlung/PDF-Dateien/tabII6.pdf (accessed 14 January 2013).

United Nations Environment Programme (UNEP) (2011) 'Towards a Green Economy: Pathways to Sustainable Development and Poverty Eradication', www.unep.org/green-economy/GreenEconomyReport/tabid/29846/Default.aspx (accessed 27 January 2013).

Vivid Economics (2012) 'Carbon Taxation and Fiscal Consolidation: The Potential of Carbon Pricing to Reduce Europe's Fiscal Deficits', London.

World Bank (2005) *Environmental Fiscal Reform What Should be Done and How to Achieve it*, Washington, DC: World Bank.

# 8    The decline and revival of environmentally related taxation in Europe

*Mikael Skou Andersen*

## Introduction

In May to June 1995 Eurobarometer as part of its regular four-year survey on 'The Europeans and the environment' posed several questions regarding environmental tax reform to European citizens. These questions were addressed to respondents in the then 15 member states of the European Union and the survey findings in a way reflect the high hopes and optimism that prevailed at the time in the mid-1990s for environmentally related taxation in Europe.[1]

In all member states at least 75 per cent of those surveyed would agree with the basic idea of applying environmental taxes, in fact support was above 80 per cent in all but three member states. The highest level of support was recorded in Spain, Greece, Portugal and Ireland, while the Belgians and Germans were the least supportive (see Figure 8.1a).

It is rare for any category of taxes to achieve this level of popularity, yet when pressed about the dilemma that environmental taxes might involve some sacrifice for economic growth, a majority of the surveyed maintained their support for this then quite innovative fiscal instrument, now with the Nordic countries and Netherlands topping the list (see Figure 8.1b).

The questionnaire also presented the idea of an environmental tax reform, involving a shift of the tax burden in line with what had been proposed in the Delors Whitepaper (European Commission, 1993). The option of reducing income taxes and social security contributions in return for new environmentally related taxes was here tested with the surveyed. Even though the question was posed in such a way as to emphasise that the tax shift would take place at member state level, overall support by no means declined as compared to the more generic question about environmental taxes – about 80 per cent agreed with the option of a revenue-neutral environmental tax reform, with few differences among member states (see Figure 8.1c).

Finally one question reflected a more sophisticated use of environmental taxation and presented the option of introducing taxes on packaging that would match the specific environmental burdens of the various types of packaging. This option was wholeheartedly endorsed by the surveyed, as close to 90 per cent welcomed such environmentally related taxes reflecting more closely the costs and burdens involved (see Figure 8.1d).

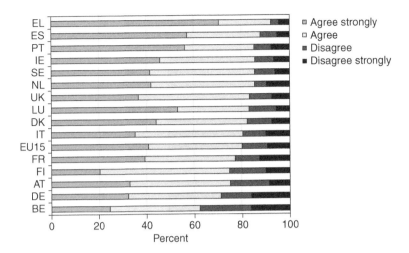

*Figure 8.1a* 'A possible means of slowing down the damaging effect of human lifestyle on the environment would be to introduce environmental taxes. Do you agree with this idea?' (source: Eurobarometer cf. European Commission, 1995).

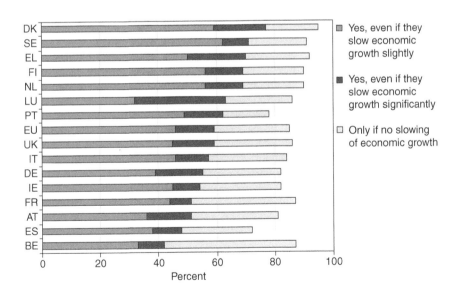

*Figure 8.1b* 'Do you think such taxes should be introduced even if they slow down economic growth?' (source: Eurobarometer cf. European Commission, 1995).

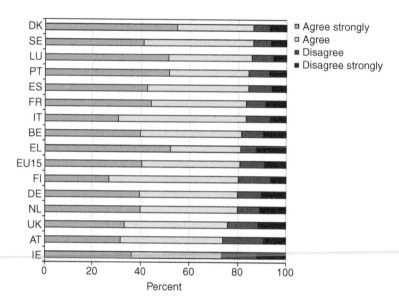

*Figure 8.1c* 'Imagine that in our country it was proposed to cut income taxes or social security contributions, but an equivalent amount of taxes would be put on goods and processes which damage the environment' (source: Eurobarometer cf. European Commission, 1995).

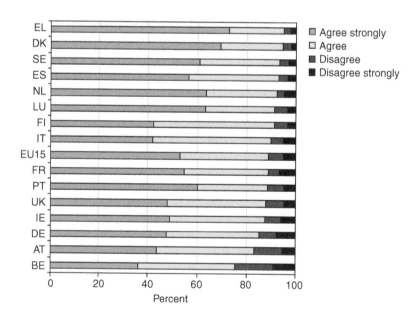

*Figure 8.1d* 'It has been proposed to increase taxes on product packaging that pollutes the environment and decrease taxes on environmentally friendly packs' (source: Eurobarometer cf. European Commission, 1995).

In the second half of the 1990s several EU member states introduced and increased environmentally related taxes, some of them as part of actual environmental tax reforms. This is reflected in Eurostat taxation data that shows an overall peak in the share of environmentally related taxes some years later, in 1999 (2.82 per cent in relation to GDP, respectively 6.94 per cent of TSC (total tax revenue, including social security contributions, SSC)). However, from that time and up to the year 2008, when the financial crisis appeared, environmentally related taxes gradually declined in significance. In some member states, such as Italy and Portugal, the decline was very pronounced and represented a relative loss in revenue significance of about 30 per cent. Under the turmoil of the crisis, with declines in GDP, the share has increased slightly again in the EU, but the 15 'old' member states generally remain well below the level of significance back in 1995 when the Eurobarometer survey was carried out (Eurostat, 2012).

This chapter reports on recent trends in environmentally related taxation in the European Union member states and explores what underlying trends can explain the paradox that a tax instrument which was in fact relatively popular with the electorate came to decline in revenue significance. This more factual analysis is then taken as a starting point for a closing discussion and analysis of the prevailing trends in member state taxation and the EU harmonisation processes in the area of taxation.

## Trends in environmentally related taxation from 1995 to 2010

Eurostat traditionally has reported environmentally related tax revenues in relation to GDP, but has complemented it also by their share of total tax revenues.

When reported in relation to GDP the figures reflect the share of the economy which is appropriated for public expenditures by means of environmentally related taxes. A higher figure for environmentally related taxes in relation to GDP reflects a more substantial role for these taxes in the economy. The proportion of environmentally related taxes in relation to GDP may also be influenced by the relative size of public expenditures, which differ significantly among member states, from 48 per cent in Denmark to 27 per cent in Romania. Furthermore, because of recent fluctuations in GDP as a result of the fiscal crisis, taxation trends are becoming more difficult to distinguish when related to GDP.

When environmentally related taxes are reported as a share of total taxation (including social contributions), then it can be seen to reflect the overall role accorded to this tax base in the taxation mix chosen by member states. The overall level of taxation expresses the preferences for public expenditures as well as more short-term decisions on how to manage the economy. The share of environmentally related taxes of total taxation then reflects the relative significance of these taxes as a source of revenues.

Figures 8.2a and 8.2b provide for EU member states as of 1 January 2013 an overview of developments from 1995 to 2010 for environmentally related taxes in relation to GDP and as a share of total taxation respectively. According to Eurostat conventions the revenues are split between energy, transport and pollution and resources respectively. The category 'energy' comprises taxes on the transport fuels, notably petrol and diesel, which in fact is the predominant source of revenues. In addition taxes on other fuels and electricity are included here, including any taxes on carbon or $CO_2$ as related to energy use. The category 'transport' comprises mainly taxes on vehicles, notably registration taxes and annual user taxes. The category 'pollution and resources' comprises the wider range of taxes relating to water, waste, air pollution and water pollution, as well as specific resource taxes for aggregates etc.

When inspecting Figures 8.2a and 8.2b it is seen that environmentally related taxes declined for EU27 as a whole by about 0.35 per cent of GDP, while measured as a share of total taxes including SSC they declined by 0.7 per cent, a decline of about 13 per cent in their tax share. The general patterns mask considerable differences among member states, reflecting the fact that whereas relative reductions can be noted for most of the old member states (Netherlands and Austria are the exceptions) several of the new member states (notably Estonia, Bulgaria, Poland) have been increasing their share of environmentally related taxes.

In terms of ranking when considering environmentally related taxes as a share of GDP, in 1995 Denmark and Netherlands topped the list of the old EU15, which took seven of the top ten places, but by 2010 the next in line were five of the new member states (Slovenia, Malta, Estonia, Bulgaria and Cyprus). Other member states, which have marketed their environmental tax shifting as success stories, such as Sweden, Finland and Germany, are only further down the list.

The ranking among member states changes somewhat when considering environmentally related taxes as a share of total taxation rather than as a share of GDP. For the share of total taxes, and with the exception of the Netherlands, it is new member states which top the list in 2010, e.g. Bulgaria, Slovenia, Malta, Latvia and Estonia. This reflects the fact that environmentally related taxes play a somewhat greater role as a source of revenue in economies in transition, the reasons for this are discussed in the section below.

## Trends in real effective tax rates

It is worth seeking an explanation for the decline in environmentally related tax revenues that can be observed between 1995 and 2010. Some such decline from behavioural changes in response to environmentally related taxes may be expected, but such changes can only be separated out from the time-series by appropriate econometric analysis, which is beyond the scope of this chapter. As an alternative, analysis of the real and nominal tax rates can investigate whether policymakers in fact have been opting for lowering of environmentally related taxes. A lowering can take place through nominal reductions in the tax rates, but also more indirectly when failing to keep up with inflation or ignoring currency depreciations.

Figure 8.3 (a,b) shows the real effective tax[2] rates for petrol and diesel in a selection of six member states; three Mediterranean Eurozone countries (Italy,

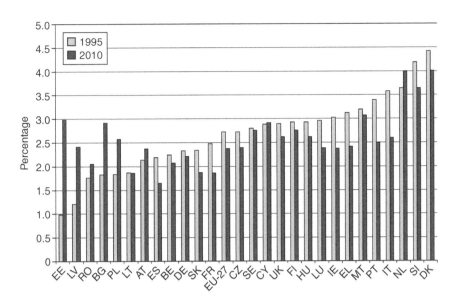

*Figure 8.2a* Environmentally related taxes in relation to GDP, ranked by their level in 1995 (source: Eurostat cf. European Commission, 2012b).

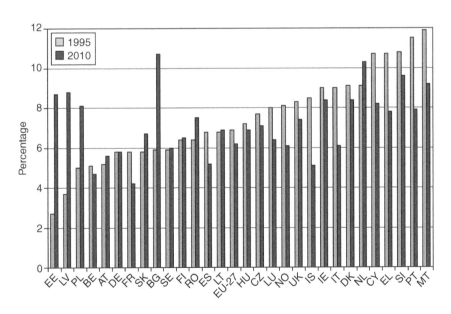

*Figure 8.2b* Environmentally related taxes as a share of total taxes (including SSC), ranked by their level in 1995 (source: Eurostat cf. European Commission, 2012b).

Portugal and Greece) and three transition economies outside the Eurozone (Poland, Estonia[3] and Bulgaria). These member states have been chosen as representing either side of the continuum, i.e. featuring either significant reductions in revenues from environmentally related taxes or significant increases.

Tax revenues from petrol and diesel taxes account for 75 per cent within the energy category of environmentally related taxes. This category provides a good basis for exploring developments in environmentally related taxes, both because it is the most significant category of environmentally related taxes, and because 95 per cent of the decline in revenues relates to this category. Focusing on this subset of environmentally related taxes also allows for an improved, if still simple, understanding of some of the general trends that were observed in the previous section and which were noted in relation to Figures 8.2a,b, in particular the different trends in transition economies versus Mediterranean Eurozone countries.

The European Union has with the Energy Taxation Directive (2003/96/EC) agreed on harmonised minimum tax rates for motor fuels and certain other energy carriers including electricity. These minimum rates have been phased in gradually for the individual member states, so as to gradually narrow the band of tax rates originally in place. By the year 2013, ten years after the adoption, the minimum rates for motor fuels have been completely harmonised for all 27 member states in the European Union.

Figure 8.3a on petrol shows how there was convergence towards the minimum rate agreed with the energy taxation directive. On one hand Poland and Estonia have increased their tax rates to comply with the minimum rates, while Bulgaria remains a special case due to the years of hyper-inflation (1996–97). Italy, Portugal and – until 2009 – Greece have seen a gradual erosion of their petrol tax rates. Italy used to have the highest level of petrol taxation in Europe, mitigated by handing tourists petrol discount coupons upon entering the country. From the mid-1990s and up to 2010 the real petrol tax rate declined by more than 20 per cent in Italy.

Figure 8.3b on real diesel tax rates further underlines the convergence that has been taking place across Europe up to 2010. Member states are now closer to the agreed EU minimum for diesel taxation; whereas for Poland, Estonia and Bulgaria this has involved a substantial increase in real tax rates, for Italy there has been a decrease comparable to the one noted for petrol. Portugal and Greece after an initial decline managed to recover the real tax rate for diesel. Diesel vehicles make up more than one-half and two-thirds of personal vehicles in Italy and Portugal respectively, but only 4 per cent in Greece, so the trends for diesel are mainly relevant in relation to the two first mentioned countries (ICCT, 2011).

An inspection of nominal tax rates show that the three transition economy countries have in fact not at any time lowered their nominal fuel tax rates, which implies that any temporary declines that can be observed in Figures 8.3a,b with the real tax rates must result from temporary inflation effects or currency depreciations. Because these countries suffered from high inflation rates in the early 1990s, some of them even had put in place certain

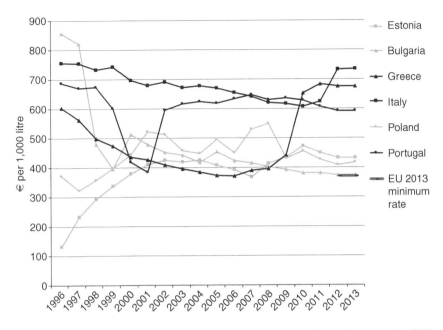

*Figure 8.3a* Real effective fuel tax rate for petrol in selected member states in 2012 prices (source: IEA, 2005–2013; OECD and EEA, 2013; REC, 1999).

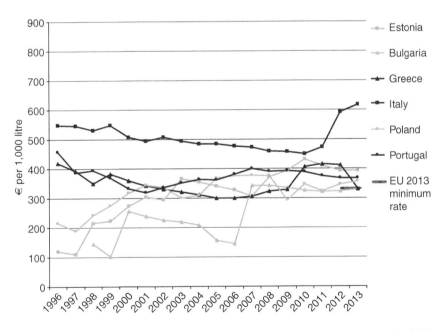

*Figure 8.3b* Real effective fuel tax rate for diesel in selected member states in 2012 prices (sources: IEA, 2005–2013; OECD and EEA, 2013; REC, 1999).

mechanisms for adjusting their environmentally related tax rates with inflation (for Estonia see UNECE, 2001: 78).

As for the Mediterranean Eurozone countries, fuel tax concessions were introduced in all three member states in 2000 in response to the fuel tax protests that followed a rise in the international oil price at this time. Italy and Portugal further lowered their fuel tax rates in 2001, while Greece lowered its petrol tax over a period of four years starting already in 1997. Still, these reductions were relatively minor (with the exception of petrol in Portugal), and in the range of 5–7 per cent, which underlines how in fact it was more likely inflation – and before 2002 also currency depreciations – that played a key role in hollowing out the real effective tax rates for petrol and diesel in the three Mediterranean Eurozone countries. Again the example of Italy is illustrative in that the nominal tax rates for petrol were not adjusted in six consecutive years from 2005 to 2010, in spite of annual inflation of about 2 per cent. Also Portugal maintained petrol and diesel tax rates practically unadjusted for six consecutive years, from 2007 to 2012, while Greece featured a five-year period with unadjusted tax levels from 2001–05.

The possible role of inflation is mirrored in the declines in real effective rates being more significant in the years prior to introducing the euro for the three Eurozone countries. Overall, from 1995–2010 Italy and Portugal experienced a decline in the real effective tax rates for the dominant fuel, diesel, of 15–18 per cent, while experiencing price inflation of 35–38 per cent. In Greece the real effective fuel tax rate declined by 20 per cent up to 2009 (2010 saw a 25 per cent increase though), while inflation was just below 30 per cent. In sum the absence of sufficient inflation adjustment can account for a great deal of the revenue erosion between 1995 and 2009/10 in these three countries.

Still, further variables need to be factored in to understand the marked decline in revenues from environmentally related taxes across the EU as a whole.

## Trends in real effective tax revenues

One difficulty with interpreting the decline in environmentally related taxes in relation to GDP is that GDP has not developed in a coherent way across the European Union. While transition economies have experienced significant increases in GDP, most of the old member states have experienced little or moderate economic growth.

Considering the real effective revenues, adjusted for inflation, can clarify the extent to which the relative revenue erosion of environmentally related tax revenues in relation to GDP at the level of EU27 can be explained by a widening gap between growth in GDP and growth in revenues (see Table 8.1).

Italy, Portugal and Greece are countries in which revenues from energy taxes declined not only relative to GDP but also in real terms – the inflation adjusted revenues were less in 2010 as compared to the beginning of the period. Poland, Estonia and Bulgaria on the other hand have experienced increases in their energy taxation related revenues that well exceed their growth in GDP.

*Table 8.1* Energy taxes: trends in real revenues over 1996–2010 in EU member states as indicated in parentheses

| *Decrease in real revenues* | *Increase, but falling behind GDP increase* | *Increase, relatively more than GDP increase* | *Increase, relatively more than twice of GDP increase* |
| --- | --- | --- | --- |
| Romania (−79%) | Sweden (4%) | Denmark (24%) | Czech Republic (98%) |
| Italy (−23%) | Belgium (5%) | Austria (37%) | Latvia (164%) |
| Hungary (−19%) | Germany (7%) | Netherlands (57%) | Malta (172%) |
| Greece (−9%) | Spain (11%) | Poland (121%) | Lithuania (318%) |
| France (−9%) | United Kingdom (14%) | | Bulgaria (319%) |
| Portugal (−5%) | Finland (18%) | | Estonia (400%) |
| Slovenia (−4%) | Luxembourg (41%) | | Cyprus (470%) |
| | Ireland (63%) | | |
| | Slovakia (65%) | | |

Source: Eurostat cf. European Commission, 2012b.

Only a handful of member states experienced an actual decrease in their energy tax revenues in constant prices; in addition to the three Mediterranean countries mentioned also France together with Slovenia, Hungary and Romania.

All the remaining member states saw an absolute increase in their energy tax revenues, but in particular for the old EU15 member states rarely sufficient to keep up with developments in GDP.

Luxembourg has benefited from a surge in fuel tank tourism after Germany's Ecological Tax Reform of 1998, helping to boost real revenues by about 40 per cent, yet its relative revenues as a share of GDP have declined. The UK is the only member state to tax diesel at the same level as petrol and the UK's fuel tax rates have in fact increased from 1995 to 2010, yet the relative revenue decline in UK is close to the EU27 average. These two cases clearly illustrate how keeping revenues up with GDP increases can be quite a challenge even in member states with increasing revenue flows.

Only three of the 'old' EU15 member states managed to increase their energy tax revenues above the level of GDP increase: Netherlands in particular but also Denmark and Austria. Here additional non-motor fuel taxes that have been introduced would appear to explain the trend.

The remaining eight member states with energy tax revenues increasing above the level of GDP belong to the group of new member states. With the exception of Malta and Cyprus these are economies in transition which featured rather low fuel tax rates in 1995 in the same way as can be seen in the figures above for Estonia, Poland and Bulgaria. Slovakia and Ireland also experienced substantial increases, but nevertheless well below their GDP growth.

For Germany, Sweden and Finland their trend is surprising, considering that these countries have introduced environmental tax reforms during the period considered which implied fuel tax increases. Also Spain and Belgium show relatively stagnant energy tax revenues. In the case of Spain an explicit lowering of the diesel tax rate to the agreed EU minimum rate seems to play a role but for the others a further and more subtle mechanism might be at play: the erosion of revenues due to the dieselisation of the vehicle fleet resulting from the tax advantages of diesel relative to petrol in most member states, because of the lower tax rates on diesel, implying a loss of revenues over time.

## Trends in vehicle fleets and implications of dieselisation for revenues

According to Ajanovic (2009) one of the most remarkable developments in passenger car transport in recent years has been the rapid gain of market shares of diesel vehicles in many European countries. Ajanovic emphasises the role of significantly lower fuel taxes for diesel in contributing to this trend. These lower tax rates are in part being justified by diesel vehicles being more energy efficient than petrol driven vehicles.

About 50 per cent of cars sold in the European Union are diesel cars, with shares above 70 per cent in member states such as France and Spain (ICCT, 2011: 39). Diesel cars have a better fuel economy, on average about 30 per cent better than petrol cars, but this improved fuel economy is not only due to higher fuel efficiency but also due to the pronounced differences in the levels of taxation for diesel and petrol. Except for the UK, all member states tax diesel more leniently than gasoline with discounts ranging up to 25–35 per cent.

Between 1995 and 2008 the number of diesel cars in the vehicle fleet is believed to have tripled from only 12 per cent to well over 35 per cent, judging from data available for ten member states (Ajanovic, 2009: 3). In France, Italy and Austria the stock is close to 60 per cent. Obviously this transformation has implications for the motor fuels sold and due to the differences in tax rates also for the revenue flows.

Ajanovic reports a 9 per cent improvement in the energy efficiency of the vehicle fleet between 1980 and 2007 of which 1 per cent can be attributed to the switch from petrol to diesel, when taking into account a rebound effect of the switch, such that the generally larger diesel cars are driving more kilometres. In contrast, and assuming only average European figures, one result of the tripled European diesel fleet appears to have been a reduction in fuel tax revenues of about 6–7 per cent. A more careful assessment taking into account the specific fuel tax rates of member states might lead to an improved understanding of the mechanisms in play in countries such as France which experienced a loss of energy tax revenues between 1995 and 2010.

Fuel tax rebates are overcompensating diesel cars, partly because diesel is also in demand from lorries that can fuel strategically where prices are most advantageous. For this reason numerous member states hesitate to raise their diesel tax above the level of their neighbours', at the same time facing difficulties with countries such as Luxembourg and Slovenia that operate with low diesel taxes only to attract international lorry traffic for fuelling.

A modest diesel tax can however be complemented by a *surplus* annual circulation tax for diesel passenger vehicles to offset lost revenues (from the tax differential between petrol and diesel) as compared to petrol cars. This approach has been practised for many years in Denmark, but the offsetting revenue is then counted under transport-related taxes and does not show up in the category of energy taxes.

A more far reaching approach would be to tax lorry diesel use according to the road sections travelled rather than at the point of fuelling. This would allow for more comparable taxation of petrol and diesel in individual member states.

## Final observations

EU GDP grew by nearly 35 per cent over 1995–2008 and reduced by approximately 3.5 per cent over 2008–2010. This represents an average annual growth of 1.8 per cent over the whole period. To maintain environmentally related taxation at the same level in relation to GDP as in 1995 real tax revenues should have grown at the same rate, but they did not.

Patterns differ considerably across Europe. New member states with economies in transition generally increased their tax take with environmentally relevant tax bases. Their revenues increased by about 50 per cent in constant prices. Old member states on the other hand largely failed to adjust motor fuel tax rates and in some cases even lowered them, notably during fuel tax protests in 2000. As a group the old member states managed to keep their real revenues relatively constant with a small increase of 4–6 per cent, ranging however from big reductions in Italy to large increases in the Netherlands, and falling well behind increases in GDP.

Presumably the Energy Taxation Directive of 2003 and the agreed minimum levels of taxation explain a great deal of the dynamics with regard to motor fuel tax rates. On one hand the new member states needed to align their much lower historical tax rates to the level agreed in the EU. On the other hand some old member states may have opted not to adjust their rates for inflation so as to fall back and over time approach more closely the minimum levels.

The consensus achieved in the Council of Ministers on the Energy Taxation Directive itself may in fact have been promoted by the fuel protests, as member states in the years 2000–02 found themselves in a 'downwards spiral' of fuel tax adjustments, affecting France and even Germany.[4] In view of the enlargement, the old member states were facing the risk of having to harmonise motor fuel tax rates at a low level, unless a compromise on minimum tax rates could be agreed. Between the Copenhagen Enlargement Summit in December 2002 and the new member states entering in 2004 the Energy Taxation Directive minimum rates were finally agreed.

These observations would seem to underline the key significance of the Energy Taxation Directive (ETD) for the overall revenue flows from environmentally related taxes in Europe. The presently stalled negotiations in the Council of Ministers are not likely to progress, unless some external pressures can reignite the interest of member states in expanding their revenue base.

In the current process of fiscal consolidation in EU member states with budgetary challenges, the European institutions have stressed the advantages of shifting the tax burden away from labour taxation and towards environmentally related and other indirect taxes. These recommendations have so far not been linked directly with the principles laid out in the proposal for ETD amendments, which not only imply a more harmonised structure for energy taxation but also that diesel will be taxed in a way more consistent with petrol and also reflecting the fuels' energy contents and associated $CO_2$ emissions.

The recent protests from Spain against requirements to increase the diesel tax rate as part of its budget consolidation process highlights the unresolved dilemmas on the taxation of motor fuels. The analysis presented here further suggests the key role that diesel plays in the gradual erosion of environmentally related tax revenues, both from the transformation in the vehicle fleets, and, more indirectly, as reflected in the propensity of certain member states not to adjust their fuel tax rates according to inflationary developments.

The most recent survey on the attitudes of European citizens to the environment was published in 2012 and offered only a rather simplistic opinion poll. It would be refreshing with a new special Eurobarometer survey to check on attitudes of European citizens to environmentally related taxes and some of the dilemmas associated with shifts in the tax burden.

## Notes

1 Eurostat has reported the survey results from the specific questions on environmental tax reform only in a rather general way (European Commission, 1995). The detailed results cited here for the first time have been retrieved from the GESIS data server at the Leibnitz Institute for Social Sciences: http://zacat.gesis.org/webview/index.jsp?object=http://134.95.45.58:80/obj/fStudy/ZA2639.
2 Net of inflation as well as currency depreciations (euro introduced in 2002).
3 Estonia joined the Eurozone only in 2011.
4 Germany in the end opted to raise a commuter tax allowance to ease fuel price protests (*ENDS Daily*, 25 September 2000).

## References

Ajanovic, A., 2009, 'The effects of dieselization of the European passenger car fleet on energy consumption and $CO_2$ emissions', Vienna University of Technology, Vienna, mimeo.

European Commission, 1993, 'Whitepaper on growth, competitiveness and employment: The challenges and ways forward into the 21st century', COM(93)/700 parts A and B, Brussels.

European Commission, 1995, 'Europeans and the environment', Survey conducted in the context of the Eurobarometer 43.1 bis, Brussels.

European Commission, 2012a, 'Taxation trends in the European Union', Brussels.

European Commission, 2012b, 'Environment and energy', Eurostat statistics in focus, no. 53.

ICCT (International Council on Clean Transportation), 2011, 'European vehicle market statistics', Washington DC.

IEA (International Energy Agency), 2005–2013, 'Energy prices and taxes', Paris.

OECD and EEA, 2013, 'Database on economic instruments', Paris.

REC (Regional Environmental Center for Central and Eastern Europe), 1999, 'Source-book on economic instruments for environmental policy', Budapest.

UNECE, 2001, 'Environmental performance reviews: Estonia second review', New York and Geneva.

# 9 Green fiscal commissions in Europe

*Paul Ekins and Ben Shaw*

## Introduction

This chapter presents the findings from a review of past and present green tax (or fiscal) commissions in Europe, and from their experience distils suggestions for the effective establishment and operation of such commissions in the future. A brief early review of experience with green tax commissions is found in Schlegelmilch (1997).

By 'green tax commission' (GTC) we mean a body that has been specifically set up to debate, develop and analyse possible options for environmental taxes or tax reform (ETR) with members 'commissioned' from multiple organisations. We exclude existing bodies that have investigated ETR within their broader work, such as standing governmental committees, unless this work has formed a distinct and ongoing project with 'commissioners' being involved in it. We have included bodies formed solely of government representatives where membership was drawn from across government.

This chapter summarises the various commissions' form, function, methods and recommendations. More detailed information on these aspects is given in the scoping paper from which much of this chapter is drawn (Shaw and Venn 2007), which also provides contextual information relevant to the formation and work of the commissions. This is available on request from the Policy Studies Institute. In this chapter we have updated the situation since 2007 and have also included comparable information on the setting up and running of the UK Green Fiscal Commission based on the authors' involvement in it.

The chapter considers GTCs that have existed in ten European countries: Austria, Belgium, Denmark, France, Germany, Ireland, Netherlands, Norway, Sweden and the UK. GTCs have also existed in Japan and the USA but these were outside the scope of this chapter, which focuses on the European experience.

### Research approach and limitations

This chapter is based on information collated from a wide range of sources. This includes information produced by the individual commissions, from a literature search covering academic journals and books and other non-peer reviewed sources.

A preliminary literature search was conducted to identify countries that had formed green tax commissions. A research template was used to structure comparable information on a wide range of aspects of each of the commissions identified: objectives, outputs, target audiences, form (size, leadership, secretariat etc.), the context to the commission's work and information on the impact of the commission. The completed templates are included in Shaw and Venn (2007).

The templates were completed as fully as possible but information was not available for all the commissions in all categories. Many of the commissions were formed and conducted their work in the early 1990s in languages other than English. As such primary sources in English do not appear to exist for the older commissions, although availability improves for more recent commissions. A wide range of secondary sources exist which mention green tax commissions in more or less detail. However, the primary focus of these studies tends to be the broader topic of environmental tax reform, rather than specifically the role of green tax commissions. As already noted, this chapter is largely based on Shaw and Venn (2007) but we have updated the material to reflect developments since 2007.

The Annex to this chapter provides an overview summary table of the commissions examined, updated from the 2007 version.

## General findings

All references to 'commissions' in what follows only relates to those that have been studied in this research.

### *Commissions were mainly formed in the 1990s*

The majority of the green tax commissions were formed and completed their work in the 1990s in a very different political and economic context from the present. Many of the commissions were created to formalise and develop broader policy and/or public debates that had been taking place in the late 1980s and early 1990s in each of the countries concerned, either on tax reform generally, or, more specifically, on environmental tax reform. Commissions have been active after 2000 in France, Ireland, the Netherlands and the UK.

The commissions after 2000 operated in a very different environment from the earlier commissions, related to a combination of economic downturn and political shifts in the European countries with later GTCs, which shifted the focus of political debate to more politically immediate economic issues rather than broader environmental considerations. Economic stability and long-term growth, combined with political consensus on the challenge of climate change, if not the response, are all likely to facilitate political discussion of broader environmental issues, which is often seen as discretionary in times of economic downturn.

## Almost exclusively formed by government

In total, 17 out of 19 of the European green tax commissions studied were formed by government. Green Budget Germany and the UK Green Fiscal Commission are the exceptions, the former created as an NGO in 1994 and the latter initiated and coordinated by a research institute (Policy Studies Institute). Green Budget Germany continues its work to the present day in contrast to many of the government-formed and -led commissions formed at the same time.

## Commission membership

The source of commission members varies widely across the commissions, from inter-ministerial working groups formed entirely of civil servants (Ireland, Denmark), to Parliamentary Committees with expert input (Sweden), to groups formed of external experts (Belgium) to groups with fuller stakeholder representation (Austria, France, Germany, the Netherlands and Norway). The size of commissions also varies with Norway having the largest commission membership we have identified, at 116 members, but the typical size is much smaller than this.

## Objectives and function

The objectives of the commissions all tend towards the general: for example, to investigate the impact of ETR on the economy and environment, or the development of proposals for a business energy tax. The scope of all the commissions studied covers environmental taxes and recycling of consequent revenues. Beyond this most commissions have considered ETR in the context of broader reform, but less than half of the commissions consider either damaging subsidies or other damaging fiscal effects.

The main functions of the commissions include:

1   developing and appraising new ETR options for consideration by government;
2   evaluating existing (environmental or non-environmental) measures in place;
3   developing principles or guidance for implementation of ETR;
4   improving implementation of measures;
5   gaining stakeholder involvement and buy-in to proposals.

Most of the commissions focus on the first three of these functions. Existing measures and new proposals were variously considered in terms of environment, social, economic, competitiveness and revenue raising impact. The methods used to appraise or evaluate new or existing measures are often not given in detail. However, a number of the commissions use general equilibrium economic models to assess the potential effect of proposals and measures. These include the commissions in Denmark, Norway and Sweden. The Dutch commissions also appear to have used rigorous evaluation techniques. In Austria and UK modelling was conducted by a body outside the commission. Other commissions may have also used modelling but no evidence was seen of this.

It is striking in the information gathered on GTCs that there is a lack of explicit consideration of stakeholder views or public support for ETR generally, or on the proposals that the various commissions have developed. This is not to say that this information does not exist, just that it isn't given high prominence in the sources examined. All except three of the commissions have external input from either experts or stakeholders. This lack of consideration of public support for ETR led the UK Green Fiscal Commission to include opinion polling, and deliberative days were held to canvass public views on ETR in its work. More broadly, for the other commissions, it is not made clear how stakeholder input was integrated into their work, or what impact it had on it.

The commissions' focus was mainly on policy development and the likely impact of ETR options based on review of existing evidence, *ex ante* appraisal and modelling. Developing instruments that are supported by economic theory or modelling is important but overlooks the fact that public opinion can be as much a barrier to the development of ETR as theoretical considerations. The PETRAS study (PETRAS 2002; Dresner *et al.* 2006) is an important study in this area. Essentially, the work indicated that the public and business have concerns about the development of ETR. These concerns are important but not insuperable and can be addressed in the design and implementation of ETR and the institutions that develop it. Institutional independence from government is considered important not only for the development of ETR but also its implementation.

In the Netherlands there was a clear separation between the work of the commission to develop possible proposals and the stakeholder process used to agree which of these could be taken forward in the National Environmental Policy Plan (NEPP) process.

A summary of the recommendations made by the commissions is presented following a brief description of the main activities, outcomes and lessons from each of the country commissions studied.

## Country commissions

Information is given below on green tax commissions in the following countries: Austria, Belgium, Denmark, France, Germany, Ireland, Netherlands, Norway, Sweden and the UK. Each country section has an introductory narrative, which covers the commission and the context to its operation and is followed by the main conclusions and lessons drawn from each commission. Further details for the commissions studied up until 2007 are given in Shaw and Venn (2007), which includes a completed research template for each country that collates the information found under specific research questions.

All pre-Euro (€) currency conversions into sterling (£) in this chapter are based on historic exchange rates and were sourced from www.oanda.com. No attempt has been made to convert historic values into current values. In August 2013 £1 was around €1.16.

### Austria

There have been at least two GTCs in Austria. These are discussed below. A third is suggested in the literature as being formed by the Green Party. However, it is not clear whether this latter one was a commission with members involving themselves in an ongoing programme of work, or more campaign-focused activity feeding into the first commission.

The first, the Austrian $CO_2$ Commission was formed in 1993 in response to earlier (as early as 1983) national debates on energy taxation (Ekins and Speck 2000). The Commission was formed by the Austrian government and was made up of 'among others, all the ministries concerned by the matter [of climate change]' (UNFCCC 1995). At the same time ETR was being championed by the Green Party, unions and other environmental NGOs although it was fiercely oppose by industrial actors (Wurzel *et al.* 2003). The Commission debated the levels of energy taxation, labour taxation and energy efficiency in households and industry. Taxes proposed included mineral oil, fossil fuel and electricity generation taxes. It was proposed that tax rates be increased 7 per cent per annum with revenues being spent on 'ecological measures' and labour tax reductions.

Only some of these proposals were implemented. In July 1996 taxes on natural gas and electricity were introduced. These raised ATS3 billion (£150 million) in 1996 and ATS7.5 billion (£375 million) in 1997 (Hoerner and Bosquet 2001). However, the proposals from the Commission on the use of tax revenue were not implemented, nor were revenues used to finance a shift in the tax base away from labour and towards environmental harm. This was largely due to the Commission failing to receive wider support from the Social Democratic–Conservative (SPÖ/ÖVP) coalition government. The discussions on ETR were overshadowed and eventually silenced by a domestic budget deficit crisis.

In 1997/1998 the ETR debate resurfaced as the Austrian government established a Commission on Tax Reform, whose focus was explicitly aimed at looking into environmental issues. The Commission rejected arguments presented by the highly regarded Austrian Institute for Economic Research that environmental tax reform would create additional jobs rather than endanger the competitiveness of Austrian industry (Wurzel *et al.* 2003). The environment ministry was only invited as an expert and NGOs were kept outside the process. After roughly one-and-a-half years of discussions and negotiations, however, both Chancellor Klima (Social Democrats) and Vice-Chancellor Schüssel (Conservatives – now/new chancellor) decided to put an end to discussions about the adoption of a national ecological tax reform by ruling it out (Wurzel *et al.* 2003). During the Conservative–Right Wing government which was in power from 1999 environmental tax reform became a 'taboo subject' (Wurzel *et al.* 2003).

While there have been developments in environmental taxation in Austria since the work of the Commission on Tax Reform the measures have been relatively minor and have not been developed in the context of an explicit programme of environmental tax reform.

*Main conclusions and lessons*

Both Austrian commissions were formed, and then disbanded due to the low level of political interest in ETR. The first commission was successful in terms of developing proposals but there was limited implementation of these for political reasons. The second commission again failed due to the external political environment. This highlights the importance of consideration by commissions of the politics and public and sectoral perspectives on ETR. However, most commissions will have limited influence over the broad economic and political trends they operate within and will, instead, have to work within them.

### Belgium

The Belgian government imposed energy taxes on gasoline, heating oil and electricity in 1993. The receipts were recycled in the form of a 1.5 per cent reduction in employers' social security contributions paid by manufacturing firms, whose export competitiveness had been hurt by the recent devaluation of several European currencies. These new taxes, which raised the price of a litre of gasoline by BEF0.65 (£0.02), were not applied to industry so consumers bore the full burden (Hoerner and Bosquet 2001). However, the cut in employers' social security contributions occurred first. It was designed as a fiscal measure supporting the macroeconomic objective of export promotion. The foregone revenue then had to be found elsewhere. The government did not invoke environmental reasons for imposing the energy tax. Instead it was preoccupied with filling a budgetary gap. This placing of fiscal priorities before environmental considerations led some to believe the energy tax cannot be considered an ETR in the strictest sense (Hoerner and Bosquet 2001).

The Belgian Green Tax Commission was not set up until 1993 and ran till 2000. Interestingly, the commission was formulated after eco-tax legislation and energy taxation changes had been implemented. This stemmed from the fact that the 1993 eco-tax legislation was a compromise born out of agreements made between a number of political parties. Since the eco-tax measures were not well designed the commission was established almost immediately to redesign the taxes (OECD 2001). The eco-taxes aimed to change consumer behaviour towards purchasing goods with a lower environmental impact, mainly in respect of waste management.

Changes to the existing ETR measures were implemented following commission recommendations on a number of products. The tax rate was purposefully set at a level to induce behaviour change rather than reflect the 'true' cost of use and disposal. Taxation applied to drinks containers, disposable razors and cameras, industrial packaging, batteries, pesticides and plant protection products, and paper.

Interestingly the government found that announcing future levies can provoke immediate action by industry, suggesting that 'cold signalling' of future intentions may be as important as consumer reaction to the tax once imposed (Martin 1997). The high rate of some of the environmental taxes introduced in the eco-tax legislation was felt to be disproportionate given the environmental impact of the goods in question. For example, the tax on disposable razors reduced sales to zero and was removed after the commission found little justification for singling out that product amongst other razor types (OECD 2001). The disposable razor tax example brings up a number of important issues around careful identification of targets for behaviour change and the ultimate intention of taxation. It also highlights the issue of weighing up different tax levels between effectiveness and reflecting the true cost of environmental impact.

### Main conclusions and lessons

The Belgian Green Tax Commission was formed retrospectively to deal with problems associated with eco-tax implementation. Its approach has a number of interesting aspects to note.

First, it is clearly focused on the practical implementation and design of eco-tax measures and there is some detailed information on these issues available. Second, it was integrated into the machinery of government and housed in the Prime Minister's office which should have ensured integration with a broader set of policies than if it had been located in a finance or environment department. Third, it appeared to have significant 'powers' to amend existing legislation and propose new measures. However, the word 'powers' may have been used inappropriately and been meant to indicate a remit. Finally, the impact or early signalling of possible taxes appears to have had an important effect on businesses' and individuals' actions.

The explicit focus on use of environmental taxes as revenue raising instruments rather than being revenue neutral is also of note.

### Denmark

During the 1980s Denmark had a tax reform debate like most OECD countries. A major tax reform cutting marginal taxes on personal income was approved in 1993 in the first year of a new centre-left government elected in January that year. This reform had been prepared by a government commission on personal income which reported in 1992 and covered a much wider scope than environmental taxes alone. The income tax reform led to a net reduction in total tax revenue, part of which was financed by an increase in revenue from environmental taxes (Mortensen and Hauch 1999).

Businesses were left generally untouched by the 1993 'green' tax reform. But in return for the support of the left-wing parties in Parliament, the text of the reform package included a Government commitment to look into further possibilities for imposing 'green' business taxes. In order to live up to its commitment the Government established an inter-ministerial commit-tee, which released its final report in early 1995.

(Klok *et al.* 2006)

The inter-ministerial committee was led by the Ministry of Finance and sat from 1993 to 1995, and was formed in response to the election of a strong coali-tion government. The environmental recommendations of the earlier personal income commission had not been implemented as the previous coalition govern-ment lacked the required political strength. However, the personal income com-mission wrote a chapter on green taxes and can be credited with introducing the idea of using green taxes to reduce taxes on income and the 'double dividend argument' into the Danish tax debate.

With the new strong coalition government in 1993 and the inter-ministerial nature of the committee, the final proposals received strong political support. However, in the legislative phase of implementing the proposals business became more involved as technical input was required which created problems. The committee recommended business environmental/energy tax reform in a number of areas. The proposals were characterised by a long-term approach to the feed in of tax reforms.

More generally environmental taxes were introduced by the government covering water consumption, leaded and unleaded petroleum, shopping bags, coal and waste disposal. The commission's recommendation to amend the business $CO_2$ tax fed into ongoing reforms. The amended $CO_2$ tax rate applied to households deviated substantially from the tax rates applied to industry and commerce, as different tax levels and reduced rates were applied for energy intensive firms. A resulting lesson was that efficiency arguments regarding the design of the ETR, made by the commission, were not important for policy makers. However, the $CO_2$ tax (on space heating) was gradually increased from DKK50/tonne $CO_2$ (£5.60/tonne) in 1995 to DKK600/tonne (£49/tonne) in 2000.

Over the period 1994–1998 labour taxes were to be reduced by 2.2 per cent of GDP and the marginal tax rates on income were cut by 10 per cent. As a coun-terpart, the capital tax base was broadened to supply additional revenue equal to 1 per cent of GDP and a broad array of green taxes was raised to provide the extra 1.2 per cent (Hoerner and Bosquet 2001).

Second, in 1996 an $SO_2$ tax was introduced on the sulphur content of fossil fuels (petroleum products, coal and natural gas) containing more than 0.05 per cent sulphur and also on the sulphur content of wood, straw, waste and other fuels burned in electricity generating plants with an output of more than one megawatt. The tax rate started at DKK10/kg of $SO_2$ and steadily increased as the $CO_2$ tax did due to the tax shift programme.

Finally an energy tax on oil, coal, gas and electricity was introduced in 1998 at a level of DKK41/GJ (£3.70/GJ) in 1998 and rose to DKK51/GJ (£4.30/GJ) in 2002. Exceptions included, *inter alia*, fuels used for energy production, air and sea transport. This meant that households and companies not VAT registered (financial sector, private health sector, public sector) would shoulder the bulk of taxes (Hoerner and Bosquet 2001).

Between 1997 and 1999 the Ministry of Finance evaluated the fiscal reforms and found that the scheme had substantial environmental effect in an economically efficient manner, and took international competitiveness into proper consideration (Danish Energy Agency 2000, p. 10). Only minor alterations were made to the tax reforms.

### Main conclusions and lessons

The two green tax commission processes in Denmark appear to have been influential. However, what is not clear is whether these successes are due to the strong political support creating conditions for the commissions to operate or vice versa.

The first, the governmental commission on personal income, set the agenda on ETR in the early 1990s and developed proposals. However, these were not adopted at the time due to a weak coalition government not having the political power to do so. On the election of a more powerful government the proposals were taken forward. Very little information is available on the form and operating methods of the commission.

The second commission, the inter-ministerial committee on energy taxation, appears to have run an effective process to develop business energy taxes, with business concerns around competitiveness an issue to be explicitly addressed by the commission.

The two commissions focused on developing the debate in different ways. The first appears to have developed the high level case for taxes and their relationship to broader economic and fiscal policy – although it seems to have resulted in specific environmental tax proposals as well. The second focused more on the development of specific proposals and issues around moving towards implementation of the taxes proposed with stakeholder support.

### France

The development of environmental taxes in France to 2000 is summarised by Deroubaix (2006). Following a long history of negotiations with industry, resulting in voluntary agreements on energy consumption which produced few results in terms of reduction of greenhouse gas emissions, the idea of an energy tax became increasingly attractive for many French decision makers. An ecological/ environmental tax reform (ETR) was implemented in France from 1999. It was known as 'Taxe Générale sur les Activités Polluantes' or TGAP. Its purpose was to unify and to simplify a set of several existing fees (on industrial and domestic wastes, atmospheric pollution, oil and noise) and to implement new taxes (on detergents, gravel extraction and energy). The revenue of TGAP was used to fund labour tax reduction. This labour tax reduction linked with the shortening and reorganisation of working time (the so-called '35 hours programme') ETR could have been one of the major political decisions and successes of the left-wing coalition government. Instead it became one of its major failures as the Constitutional Court decided to terminate the energy tax project in December 2000. The fuel protests of 2000 ended any immediate hope of resolving the conflicts and implementing ETR.

Subsequent research into the factors accounting for this outcome found that decision makers lacked crucial information about public and business opinions and, second, there were conflicts between the relevant administrations (Deroubaix, 2006).

In early 2006 a group called 'Outils Économiques et Développement Durable' was set up by both the Ministry of Finance (MINEFI) and the Ministry of Ecology and Sustainable Development (MEDD). The initiative for the establishment of a GTC came from an OECD review of environmental performance of France which, *inter alia*, recommended the creation of a GTC. However, the French 'Groupe de travail' is not purely a GTC as it covers other types of economic instruments including tradable permits. The group comprised 35 members including representatives from government departments (environment, finance and industry, agriculture, equipment), MPs, representatives from local communities, private industry, and environmental NGOs and 'experts indépendants' (including academics). The group was chaired by Jean-Pierre Landau, from the 'Banque de France' and was served by a secretariat of staff members of the Ministries of Finance and Ecology.

The working group published a report of their findings in July 2007 (Landau 2007) on the use and reform of economic instruments in the fight against climate change, biodiversity loss and road congestion. While primarily focusing on taxation and tradable permits (and their corresponding strengths and weaknesses), the report strongly advocated the use of a wide mix of complementary instruments, additionally calling for the elimination of harmful subsidies. The group concluded that the level of intervention, distributional effects on income, design with particular consideration for the long term, and the preservation of both economic competition and the concept of neutrality, were all key issues for the design of effective economic instruments for sustainable development.

Senit (2012) outlines the process of French government deliberations from 2007:

> After the main candidates to the presidential elections signed the Ecological Pact in January 2007, which aimed to place ecology at the centre of the political agenda, the newly elected President Nicolas Sarkozy implemented a deliberative process during the summer of 2007, called the Grenelle Environnement, the objective of which was to define the key points of the future French environmental public policy. In October, during the final negotiation

round tables, Nicolas Sarkozy agreed to study the implementation of a "carbon energy contribution", presumably the Landau Report. A year later, (2008) the oil price reached records of nearly US$140 per barrel and the French Government announced the postponement of the carbon tax reform.

(Senit 2012)

The French government established a further commission of experts in July 2009 with the aim of developing future economic instruments, in particular a carbon tax, focusing thereby on the tax base and rate as well as on the recycling mechanism (Senit 2012). The outcome of the commission was not fully transposed into the finance bill of December 2009 which laid out the conditions for establishing the carbon tax which was planned to take effect from January 2010. The experts proposed a tax rate of €32 per tonne of $CO_2$ which should gradually be increased reaching €100 per tonne of $CO_2$ in 2030. Instead the tax rate was set at €17 per tonne $CO_2$ levied on the use of oil, gas and coal in the finance bill for 2010. However, the finance bill was taken to the Constitutional Court stating that several exceptions as well as the recycling mechanism for households would be unconstitutional. The court allowed an objection and therefore the carbon tax project was not implemented in January 2010. The then French government began to revise the carbon tax shortly afterwards but then withdrew the whole carbon tax project after the ruling party was heavily defeated at a regional election.

In late 2012 yet another commission or 'Comité pour la fiscalité écologique' was set up to progress environmental tax reform in France, launched by the French Minister for Environment, Sustainable Development and Energy, Delphine Batho (Batho 2012), but in partnership with the Finance Ministry and to be chaired by Professor Christian de Perthuis.

Ichay (n.d.) reports that the committee was to consist of 40 members including elected officials, employees, representatives of the employers, local authorities, associations and non-governmental organizations and would also benefit from groups of experts. The aim of the committee is twofold. In the short term it will develop proposals that may be integrated into the finance bill for 2014. In the longer term, the aim is to develop thinking on the development of environmental taxation in the context of an overall reform of tax and social security deductions (Ichay n.d.).

The Committee's initial recommendation, made in March 2013, to reduce the tax advantage diesel has over petrol seemed to prompt divergent responses from different ministries (*ENDS Europe* 2013) suggesting that the long-running debate on the role of ETR in France is unlikely to be resolved in the short term.

*Main conclusions and lessons*

Obviously, the fact that TGAP and successive initiatives were shown to be unconstitutional is a fundamental problem but the failure of the process indicates the more general issue of poor stakeholder relations. Senit (2012) discusses the French context to the development of carbon taxation and highlights the gap between the academic and expert proposals developed and the political context they are to be delivered in, including factors such as electoral incentives, policy-makers' preferences, institutional rules and partisan ideologies. Senit also highlights the importance of public opinion in delivery of ETR in France including the need for a practical revenue recycling mechanism to be developed which is distributionally fair and a 'marketing strategy' to better inform the public with the importance of independent and reliable sources of information being highlighted.

### *Germany*

Green Budget Germany (GBG) was established in 1994 as a single issue NGO to support the development of an Ecological Tax Reform (ETR) and Emissions Trading (ET) in Germany. It has members drawn from industry, business, the research community, politicians from all German democratic parties and journalists, and has campaigned for ETR in Germany since the mid 1990s. In 1994, 1997 and 1998 the GBG published three memoranda on ETR. GBG members contributed to the debate in many articles in the press and were involved several times in hearings of the German Bundestag as experts. GBG also organised many events on ETR.

At the time of its formation, Green Budget Germany's focus was on the introduction of an environmental tax reform in Germany. It considers this goal to have been realised to a large extent and it has extended its focus to include all elements of market-oriented eco-fiscal policy: environmental taxation, emission trading, removal of environmentally harmful subsidies, promotion of renewable energies and green growth (Green Budget Germany, 2013).

The German government implemented ETR in 1999. The ETR was brought in through legislation passed through the German Bundestag. The 'Act on the Introduction of the Ecological Tax Reform' increased the price of energy (motor fuel, heating oil, electricity and natural gas) from 1 April 1999.

Labour in Germany is relatively expensive due to high non-wage labour costs, in particular social insurance contributions, which are amongst the highest in Europe. This is perceived to have a negative impact on Germany's competitiveness and to contribute to unemployment. A key notion behind Germany's reforms was therefore aimed at shifting the tax burden from labour costs towards environmental factors.

Phase one of the ecological tax reform in Germany ended on 31 December 2003. Since 1 April 1999, taxes on petrol and diesel, electricity, heating oil and natural gas have been increased in five stages, and the bulk of the tax revenue generated used to reduce pension insurance contributions. Approximately 88 per cent of ETR revenue was used to increase federal subsidy to statutory pension insurance institutions. Over this period, the total volume of energy taxes rose from €34.1 billion in 1998 to around €52.7 billion – an increase of 55 per cent. This is in a context of an overall 4.2 per cent reduction in the overall tax ratio[1] over the same period.

The positive effects of the ETR were highlighted by the Federal Environmental Bureau (Umweltbundesamt) in early 2002 when it stated that by the end of that year, its projections showed that ETR would have reduced $CO_2$ emissions by more than seven million tonnes while at the same time creating almost 60,000 new jobs. The ETR, GBG claims, has helped to influence changes in behaviour including trends of declining fuel consumption in road traffic and increases in passenger numbers on public transport. The promise of revenue neutrality has 'largely been met'.

The Act on the Further Development of the Ecological Tax Reform entered into force on 1 January 2003. This dismantled environmentally harmful tax reductions and adapted taxes on natural and fluid gas and on heavy heating oil. Measures included in the Act include, *inter alia*, increasing eco-tax rates for manufacturing industry, agriculture and forestry, reducing the tax reimbursements to energy-intensive manufacturers and encouraging natural gas deployment in the transport sectors. However, it seems that the German energy tax escalator was halted in mid-2003. GBG was at the forefront of criticism at the time and has in 2006 called again for the escalator to be restarted (*ENDS Europe* 2006a).

### Main conclusions and lessons

Green Budget Germany is interesting as it is a commission formed outside government. Its independence and continuity appear to be its main strengths, which need to be balanced against ensuring its work and findings are fed into the policy process, or alternatively influence policy indirectly through the pressure it creates for reform. However, it has developed broad stakeholder buy-in and is also respected enough to be an expert witness to the German Bundestag which suggest it is striking the right balance here, and that Green Budget Germany is a valued body performing a useful role in the German debate on ETR. However, without further investigation, it is hard to say how effective a body it is in driving forward ETR, and now wider progress on market based instruments for environmental policy, in Germany. Green Budget Germany was also the main driver behind the creation of Green Budget Europe, which advocates for ETR at an EU level.

### Ireland

The Irish Environmental Taxation Group (ETG) was set up by the Irish Finance Minister in 1996, having a membership drawn from six government departments including the then Department for Environment, Department of Enterprise and Employment, Department of Transport, Department of Energy and Communications and the Office of the Revenue Commissioners. The ETG was led by the Ministry of Finance and was a sub-group of the cross-departmental Tax Strategy Group (TSG) which develops recommendations on Irish budget measures (Department of Finance 1999).

The aims of the ETG were 'to examine the strategic impact of taxation on environmental policy and to bring forward specific tax measures for the 1997 Budget which will have due regard to competitiveness considerations' (Department of Finance 1999). In 1999 the ETG met seven times and developed proposals on diesel tax rebates, vehicle registration taxes, VAT on electricity, plastic bag tax, low sulphur diesel and a framework on GHG taxes (Department of Finance 1999). Although referred to positively in a 1998 speech by the Finance Minister (quoted in Department of Finance 1999), there was limited implementation of energy/ $CO_2$ taxes in Ireland and discussions on possibilities continued in the ETG which reported to the Tax Strategy Group in 2002.

In 2002 the ETG proposed an excise-type tax on all fossil fuels across all sectors, directly related to the amount of $CO_2$ emitted from their combustion. An initial tax rate of €7.50 per tonne of $CO_2$ was proposed to be introduced from 2003 to provide an adequate environmental and economic signal with the rate increasing to €20 per tonne over a period of three to four years (Department of Finance 2002).

These proposals led to the Minister for Finance publishing a carbon energy tax consultation paper on 31 July 2003 (details of the process are given in Department of Finance 2004a). Following the consultation it was announced in September 2004 that plans for an Irish carbon tax had been abandoned and that the Irish National Climate Change Strategy would be pursued by an intensification of non-tax measures (Department of Finance 2004b).

The most notable environmental tax introduced in Ireland is the plastic bag tax which was implemented in 2002. Although the plastic bag tax met initial resistance from manufacturers, the tax has proven highly effective in reducing usage and visual litter and is so popular with the public that it is felt it would be politically damaging to remove it (Convery *et al.* 2005). The Irish government carefully considered a number of options before deciding on a bag tax set at €0.15 per bag. 'The tax rate was not based on an estimate of the value of the negative environmental impacts caused by the bags, but instead at a level believed to be sufficiently high to trigger a major behavioural change' (OECD 2006, p. 123). However, while this suggests a successful measure, its effectiveness has been questioned. For example, following an extended Environmental Impact Assessment published in 2005 (Scottish Executive 2005) the Scottish Executive withdrew its support for implementing a plastic bags tax in Scotland, favouring a voluntary code approach.

In 2008 the Irish government established the 'Commission on Taxation' to 'review the structure, efficiency and appropriateness of the Irish taxation system' (Commission on Taxation 2009, p. 35). In setting up the commission, the then Minister for Finance

> indicated that its work would help establish the framework within which tax policy would be set for the next decade at least, and that it was important that it take a strategic, considered and balanced perspective that recognised the evolving challenges ahead.
>
> (Commission on Taxation 2009, p. 35)

The remit of the commission was broader than green fiscal reform and was conducted in the context of the Irish government's commitments on economic competitiveness and taxation. These included commitments to keep the overall tax burden low and in particular guarantee the 12.5 per cent corporation tax rate, but also to 'introduce measures to further lower carbon emissions and to phase in on a revenue neutral basis appropriate fiscal measures including a carbon levy over the lifetime of the Government' (Commission on Taxation 2009, p. 35).

Given the short timescale for its work (March 2008 to July 2009) and broad remit the commission split into seven subgroups including one addressing 'fiscal measures to protect and enhance the environment including the introduction of a carbon tax' (Commission on Taxation 2009, p. 39). A total of 110 meetings of these subgroups were held in addition to 32 commission plenary meetings. The commission utilised a range of methods to get broad stakeholder input including requesting submissions to the commission with the opportunity to submit material publicised in the national media, oral hearings from select witnesses and meetings with a wider range of stakeholders.

The commission reported in 2009 in a substantial (561 pages) and broad ranging report covering all aspects of taxation and the taxation system. The commission's recommendations proposed a portfolio of environmental taxes including

> a carbon tax, metered water charges, waste charges, road charges, not taking gains from EU ETS allowances, accelerated capital allowances for energy efficient equipment, favourable VAT rates for environmental goods and the offset of R&D tax credits against employer PRSI costs.
>
> (Commission on Taxation 2009, pp. 369–370)

These measures were considered 'key to Ireland's prospects of success in this emerging world' of growth in the environmental goods and services sector (Commission on Taxation 2009, p. 370)

The introduction of a carbon tax was announced in the 2010 Budget (Department of Finance 2009). The tax rate was set at €15 per tonne of $CO_2$ and was applied to petrol and diesel with immediate effect from December 2009 and to non-transport fuels (i.e. kerosene, gas oil, liquid petroleum gas, fuel oil and natural gas) from May 2010. The revenues of the carbon tax are not earmarked or recycled back to tax payers but instead used for the fiscal consolidation process, i.e. reducing the budget deficit.

The carbon tax was increased in the 2012 budget as planned from €15 per tonne $CO_2$ to €20 per tonne $CO_2$. The increases applied to petrol and diesel from 6 December 2011 and to other energy products, such as LPG, fuel oil and natural gas, from 1 May 2012. It is expected that this increase will raise €109 million in 2012. The reason for delaying the increasing of the carbon tax for other fuels (i.e. heating fuels) was to mitigate the impacts of this increase on home heating costs during winter. Furthermore, solid fuels, such as coal, are still exempt from the carbon tax.

The carbon tax rate was not increased in the budget 2013. However, the carbon tax will be extended as solid fuels will be taxed from May 2013 onwards (Department of Finance 2012). The initial rate is set at €10 per tonne $CO_2$ and will increase to €20 per tonne in May 2014. It is expected that this measure will generate about €6 million in 2013.

## Main conclusions and lessons

The initial processes used to explore ETR in Ireland and the commission it used do not seem to have much to recommend them as a model for other commissions. The first Irish commission was formed solely of government officials and its recommendations have neither been adopted nor done much to create a greater likelihood of implementation of ETR in Ireland. This may of course be as much a function of an unsympathetic external political environment as a flawed process. The commission may have had more success if it had focused on building a general case for ETR in Ireland rather than development of specific measures. The lesson here is to ensure that any commission's work is relevant to the level of development in the policy debate.

The one tax that has been implemented as a result of that commission's work, the plastic bag tax, has been important in terms of stimulating debate on these issues in the public and beyond Ireland amongst policy makers. As such consideration of measures of such symbolic significance and public profile may be of value in consideration of ETR.

Convery (2012) notes that the Commission for Taxation recommendations in relation to the design of carbon tax were followed and advice on its implementation largely accepted (Convery was a member of the Commission). Convery also notes the fiscally stressed nature of the Irish economy from 2008 and the drop in GDP and tax revenues associated with these. By 2012 the carbon tax was raising €400 million of $36.4 billion total tax revenues.

The successful implementation of the commission's recommendation may have been due a combination of the extreme financial conditions Ireland found itself in – taxes had to be increased on both income and environmental goods – combined with a commission process that engaged stakeholders and was independent. Revenue neutrality has usually been associated with proposals for ETR, although with this approach the public tends in general to object to an increase in environmental taxes and overlook reductions in taxes on labour or income elsewhere. It would be interesting to explore whether the overall increase in taxation in the fiscal situation Ireland found itself in increased the acceptability of the proposals.

## Netherlands

There have been three Dutch green tax commissions, the first formed in 1990, the second in 1995 and the third in 2000.[2]

There is little information available on the 1990 Dutch Green Tax Commission but Schlegelmilch (1997) indicates it came up with a proposal for a carbon/energy tax which was implemented in 1996.

The 'Van der Vaart' Commission was formed in 1995 and chaired by Jacob van der Vaart following a request by the Dutch Parliament and was implemented by the Ministry of Finance. The commission membership included experts in tax, economics and environment, and members from employers' organisations, research bodies, civil servants and former MPs. It received cooperation from several government planning agencies (Vermeend and van der Vaart 1998, p. 125).

The commission's task was to assess the overall provisions of the Dutch tax system and to propose how taxation could work more effectively for the environment. The major consideration in formulating proposals was environmental effectiveness, coupled to feasibility and political acceptability. The commission produced three reports[3] covering a wide range of environmental taxes including on transport, energy, waste, water, but also issues around implementation such as indexation to inflation and balancing rewards and penalties. With successive reports more ambitious proposals were made. By 1998, 14 per cent of total tax revenue was raised by green taxes (Vermeend and van der Vaart 1998, p. 118) with many of the recommendations from the first and second reports having been implemented.

Immediately following the third and final Van der Vaart Commission report in 1997, the Ministry of Finance produced a White Paper which incorporated recommendations from the third report and outlined a proposal to shift around an extra 0.6 per cent of GDP from income tax to environmental and energy taxes.

The objective of the Van der Vaart commission

> was solely to identify the contribution which fiscal policy can conceivably make to achieving greater environmental quality and sustainability. The success of the Commission must not be judged according to the number of positive suggestions it has been able to put forward, but rather by the degree to which it has been able to present the ideas and suggestions from within society in a systematic and substantiated manner.
>
> (Dutch Green Tax Commission 2001, p. 11)

The remit of the 2000 Dutch Green Tax Commission, set up under Jan de Waard, was wider than its predecessor and the commission investigated around 80 possible tax shifts. It was asked to

> Define new possibilities, evaluate the measures taken thus far, to identify any tax expenditure that may have a detrimental effect on the environment, to define the anticipated effects of new measures and to incorporate all these aspects in an agenda for the next government period.
>
> (Dutch Green Tax Commission 2001, p. 3)

The de Waard Commission reported early in mid-2001 to fit in with political party manifesto timetables. The Sustainable Development Commission (2006) suggests that the de Waard Commission was arguably less effective than the Van der Vaart one as many of the easier changes had already been made following the first commission

*Main conclusions and lessons*

The relationship of the commissions to the political process and policy making in the political parties is interesting. Although the commissions are chaired by the Ministry of Finance they appear to have been developed to fit in with the manifesto development process. The explicit remit to consider the political acceptability of measures, along with feasibility and effectiveness, is also of interest.

There also seems to be a separation between the commission processes to develop and evaluate proposals and the stakeholder National Environmental Policy Plan (NEPP) process which decides whether they should go forward. This approach could be used to prevent premature narrowing of options. The commissions, especially the first one, evaluated a wide range of possible measures that were fed into other (stakeholder) processes.

The wide range of options evaluated by the de Waard commission could provide a useful resource of potential measures that could be considered by other countries considering an ETR.

## Norway

There have been two green tax commissions in Norway. The first, established by the Norwegian government in 1990, proposed in 1992 'a comprehensive scheme for integrating environmental polices' with other policy areas (Moe 1999). This appears to have resulted in a series of taxes on various fossil fuels.

The second commission, the Norwegian Governmental Commission on Green Taxes was appointed by the Norwegian government in 1994, building on the previous commission (Ekins and Speck 2000). The green tax commission was formed and appointed by the government with 116 representatives from different ministries, NGOs, the academic community and social partners and industry. There were no political members of the commission. The aims were to 'investigate how to change the present tax system away from taxation of labour and towards activities that imply increased use of resources and harmful emissions in a longer term perspective, ie into the next Century' (NGTC 1996). A number of issues were considered in the scope of the investigation by the commission, including playing a pioneering role among countries in leading ETR reform, assessing effects on Norway as a large producer and exporter of energy and examining distributional effects and transitional problems generated by ETR. This investigation included a partial analysis of taxes on polluting emissions, rents from the use of natural resources and changes in environmentally harmful subsidies.

The commission recommendations were divided into principles for $CO_2$ taxes, proposals in other areas and proposals requiring further investigation. Proposals for $CO_2$ taxes include incorporating a 50 NOK (£50)/kg $CO_2$ tax on sectors that were exempt from existing charge systems, broadening the $SO_2$ charge base, harmonising taxes on petrol and diesel, introducing a charge on gas powered cars, harmonisation of purchase taxes on cars by raising charges on heavy cars and reducing existing transport subsidies to agriculture, road building, mining and the petroleum industry. Non-energy proposals included taxes on studded tyres, differentiation on the annual charge for heavy vehicles, a deposit refund system for waste, a charge on beverage packaging, a charge on final waste treatment (of about 200–300 NOK (£200–300)/tonne), a differentiated annual charge on leisure boats and an introduction of charges on cadmium, nickel, trichloroethylene and antibiotics for fish. Suggestions that needed further investigation included road charges, charges on emissions permits, further charges on aircraft noise, differentiated charges on ships and ship transports and a nature charge implying payments for environmentally harmful changes of use (Brännlund and Gren 1999).

The Norwegian government followed up on the GTC recommendations – with green tax recommendations included in the Norwegian White Paper on Climate Change. However, the minority government was not able to implement the expansion of the $CO_2$ tax which was partly voted down by the parliament in the summer of 1998. Despite these setbacks the Norwegian government implemented domestic aviation fuel tax in 1999 as a result of the commission's recommendations. In spite of opposition from airlines, it was passed by government: a key reason being the cross-party support for demonstrating their world leadership in green tax reform. Norway is still one of very few countries to have taxed aviation fuel (OECD 2006).

*Main conclusions and lessons*

The process used by the Norwegian green tax commission is a good example of a thorough and multi-stakeholder evaluation of possible ETR. A very large commission (perhaps too large to be manageable) developed measures suitable for implementation over different timescales and emphasised the need for structural change in the economy. The commission focused on the development of measures and principles to judge measures and used modelling to evaluate the potential impact of measures (as done by a number of other commissions). The uptake of proposed measures was thwarted by political developments. Bretteville and Søfting (2000) conclude that environmental taxes in Norway 'seem to be a declining political instrument' and suggest 'the idea of a green tax reform never got past the starting gate'. The green fiscal commission was formed as a political compromise between parties with the government not wanting a commission.

Another useful lesson is the way the commission presented proposals according to how much work is required before they could be implemented. This approach can be used to create a focus on doing what is possible but also the further steps that need to be taken to drive ETR forward.

*Sweden*

Prior to the establishment of the Swedish 1995 commission, Sweden's 1991 ETR was the first instance of a major, explicit shift in the tax base from traditional (labour) taxes to pollution taxes (Hoerner and Bosquet 2001). The shift in the tax base took the form of expanding value-added tax to energy and imposing an excise tax on $CO_2$ and $SO_2$, as well as raising a number of existing environmental levies. The Swedish Government Commission on Green Taxation was established in March 1995 and ran until January 1997. The 1995 commission included political elements as it was a parliamentary commission, formed from members of the Swedish Parliament but also included experts from various institutions and organisations such as trade unions, the Swedish Industry Federation, the Swedish Environment Protection Agency and universities. In brief, the commission was asked whether Sweden should continue 'greening' its tax system (Brännlund and Gren 1999).

The commission was formed as it was recognised that taxes can be a cost effective way of achieving environmental objectives, raising revenue and addressing high rates of labour taxation. The commission was tasked with a wide range of aims including, (a) to evaluate the prevailing economic instruments used in Swedish environmental policy, (b) to analyse the conditions for and effects of different kinds of revenue-neutral tax reforms and (c) to give suggestions for or against a tax reform towards more environmentally related taxes. The commission was asked to study the effects of ETR on issues including the environment, competitiveness, employment, efficiency in resource allocation and aggregate tax revenues. It was noted that in the analysis of the conditions for revenue neutral tax reforms it is important to investigate the stability of the corresponding tax bases. The commission was therefore asked to put specific emphasis on 'tax wedges', the labour supply, employment, competitiveness, structural change and the environment.

The commission concluded that Sweden had secured environmental improvements as a result of increasing the number of green economic instruments within the general reshaping of the Swedish tax system. Rather than recommend specific further tax changes the commission instead established the following principles to steer future environmental taxation policy. It argued Sweden should:

- continue to play an active leadership role in environmental policy; but a condition to pursue this strategy is that other countries do the same;
- promote joint action to combat environmental concerns through organisations such as the EU and UN;
- consider how investment, economic growth and equity would be affected by reforms and alleviate negative distributional impacts where possible, such as through tax cuts;
- the commission also recommends that revenues obtained through environmental or other taxes should in general not be designated (ring-fenced).

The study of the Green Tax Commission concluded that the theoretical 'double dividend' tax reforms 'are probably not welfare enhancing in Sweden' (Brännlund and Kristom 1997). However, taxation was implemented on energy products, carbon and sulphur. The commission noted that each dollar of tax revenue should be used where its utility is highest. A conclusion was that additional revenues from energy and environmental taxes would not leave room for any substantial general decrease of labour taxes. For this reason, revenues from a green tax reform should be used selectively. It was argued that tax cuts which will have the largest effects on unemployment should have first priority (Brännlund and Kristom 1997).

In 2004 the Swedish Environmental Protection Agency was asked to develop policy proposals for improving the effectiveness of green taxes, the results of which were intended to feed into green tax proposals from 2005–2010. The scope of that study was widened, on advice from the Environment Ministry, to address measures which were having a perverse environmental impact (Swedish Environmental Protection Agency 2004).

In late 2006, however, the new Swedish centre-right alliance coalition government called a halt to the 'green tax shift' programme. No justification other than stating that 'Priority will be given to developing effective policy instruments' was given (*ENDS Europe* 2006b). The ten-year programme had most recently called for a €390 million rise in green taxes in 2005. When many other countries halted energy tax rises in 2000 in the face of a wave of protests in the transport sector, Sweden continued introducing annual increases (*ENDS Europe* 2000). After Germany's energy tax escalator ground to a halt in 2003, Sweden has been virtually alone among the EU15 members in continuing to pursue the policy in practice as well as in rhetoric (*ENDS Europe* 2006b).

### Main conclusions and lessons

As with Norway, the Swedish commission was a thorough process which appears to have been influential. However, there was less stakeholder involvement in the process as it was formed of experts and parliamentarians. In addition, it tended to focus on principles rather than individual measures, and experienced the problem of political changes removing support for taking forward proposed measures.

### United Kingdom

The UK Green Fiscal Commission (GFC) was founded in May 2007 as an independent initiative, and was active until October 2009. It comprised 22 commissioners, all acting in a personal capacity, to review and advise on the GFC's work, who included: four MPs, three Members of the House of Lords, providing politically balanced, senior political representation, and shadow ministers; representatives from business, academics, and social and environmental NGOs; members of the Financial Services Authority (FSA) and Monetary Policy

Committee (MPC); as well as government observers from the Department of the Environment, Food and Rural Affairs (Defra) and Her Majesty's Treasury. Its Director was Paul Ekins and chairman Robert Napier who was also Chairman of the UK Homes and Communities Agency and Board of the Met Office and former Chief Executive of WWF-UK. The Secretariat, headed up by Ben Shaw, was provided by the Policy Studies Institute (PSI). Funding came from two charitable foundations.

The over-arching objective of GFC was to establish an evidence-base to support the large-scale introduction of green taxes, in order to respond effectively to global environmental challenges (especially climate change), and bring about cost-effective environmental improvements in a way that also promotes innovation in business and the creation of new industries.

The subsidiary objectives of GFC were:

1   to generate and present in an accessible form a definitive body of knowledge about every aspect of revenue-raising environmental economic instruments (hereafter simply called 'green taxes' for simplicity), that is of practical use to those who wish to introduce green taxes;
2   to increase understanding of the benefits of green taxes among policy makers, businesses, NGOs and other interested parties;
3   to raise the profile of green taxes in political and economic discussion.

From the evidence generated by GFC, and the public discussion and policy debate about this, it was intended that politicians of all major parties would include proposals for green taxes in their manifestos for the next General Election.

The activities of GFC to deliver these aims and objectives included briefing meetings for policy makers, both inside and outside Parliament; the organisation of two 'deliberative days', with 50 members of the public at each event, to explore public attitudes to and options for successful political presentation of green fiscal reform; economic modelling to understand the implications of a major programme of green fiscal reform scenarios; the production of eight Briefing Papers (available on the GFC website: www.greenfiscalcommission.org.uk/), based on longer supporting papers, a final report which presents the main findings of the commission, and some academic papers (Ekins *et al.* 2010, 2012). Other dissemination of the commission's findings occurred through meetings with stakeholders and policy makers in the UK and abroad and at party political conferences, the GFC blog and website, and articles in the media. As an indicator of the impact of GFC's work, the Coalition Agreement of the new UK government in 2010 included a commitment to 'increase the proportion of tax revenue accounted for by environmental taxes' (HMG 2010, p. 31).

*Main conclusions and lessons*

The value of forming a commission to explore the issues around green fiscal reform should not be underestimated. Careful choice of commissioners from a range of sectors and political backgrounds was instrumental in taking the debate beyond the environmental NGO/research sector where it had been traditionally located. This was important given the politically sensitive nature of the work of the commission. The range of commissioners also ensured the commission was seen to be independent of any particular sectoral, policy or political agenda. The expertise of the commissioners was also very important in shaping the direction of the work and its presentation.

The commissioners acted in an advisory rather than decision-making capacity. Had the commissioners had to formally agree every output, the process would have been significantly drawn out, and probably impossible.

The GFC programme of communications and engagement was successful in gaining media coverage and raising the profile and findings of the commission with relevant audiences. However, it is not easy to get sophisticated and complex ideas across to target audiences especially via the media and even when ideas are picked up they may be misrepresented or distorted. Simple and clear messages have to be repeated, again and again, with the detail following when there is pickup. The preparation of a clear narrative and line of argument for GFC (Briefing Paper 4) was a useful tool to focus on the essence of the commission's message in this regard.

## Overall conclusions and recommendations on green tax commissions

### Recommendations from commissions

A brief overview is given here of the nature of the recommendations made by the commissions. Further details are given in the research templates in Shaw and Venn (2007).

As would be expected, a large number recommendations for specific tax measures are made by the various commissions in areas, for example, such as energy and carbon, vehicles, products, chemicals, waste, water, etc., or on taxes that target specific sectors. The taxes evaluated by the de Waard Commission in the Netherlands is a wide ranging and agenda-setting list of measures and is of particular note in this area.

Recommendations covering issues around the design of environmental taxes or broader ETR are also given. For example, whether to hypothecate tax revenue or not, the size and stability of the tax base, the use tax revenue is put to, how to recycle revenues, whether taxes should be revenue neutral, the effectiveness and efficiency of taxes, the level of tax required to create behavioural impact, the use and rate of escalators, and exemptions. The detail of the recommendations is, as would be expected, much influenced by the country and wider economic context in which it is being considered and history of environmental policy.

A number of mentions are made of the need to look at incentives for pro-environmental behaviour through, for example, reductions in taxes or tax credits for particular activities. For example, the Van der Vaart Commission in its second report came out strongly in favour of more positive incentives (Vermeend and Van der Vaart 1998, p. 12).

Some of the commissions consider timescales over which it is feasible to implement particular taxes, for example, the Norwegian Governmental Commission on Green Taxes. This can provide a useful focus on what is possible now and what needs to happen to deliver the ETR agenda in the medium and long term. Related to this issue are recommendations that consider or highlight the need for more fundamental structural institutional changes that need to be made to the economy to deliver environmental objectives. Some commissions have outlined principles for the development of taxes rather than specific measures, for example, the commissions in the Netherlands, Norway and Sweden. One notable principle articulated by the Swedish commission highlights the need for international action on this issue and also the need to assess whether particular measures just cause the transfer of impacts to a different location. Social impacts and equity issues are also considered by various commissions.

### Lessons from previous commissions

The main lesson that can be drawn from studying previous green tax commissions is that they are very much at the mercy of the external political and economic context. Governments can decide to form commissions but they can also decide to disband them or ignore their recommendations. A multitude of factors will affect these decisions, including public and business opinion, economic performance and political shifts which may be beyond immediate political influence. Most of the commissions have been affected negatively by political developments during their existence. This problem is worse when the uptake of recommendations made by commissions is considered. However, the experience of the Irish Commission on Taxation highlights that there may be opportunities when the economic climate is poor.

Additionally, with political support commissions appear to perform a useful function in developing the debate on environmental taxes and broader environmental tax reform. Commission recommendations have been taken up to a significant, albeit variable degree, and without them the number of green taxes and ETR measures implemented would be likely to be less.

Being independent of government, as in the case of Green Budget Germany, removes some of the constraints that policy makers and politicians have in discussing future policy options but has to be tempered against the fact that it has less direct influence over policy. However, effective development of evidence and communication of this to target audiences, as with Green Budget Germany, appears to have been effective.

The function of commissions is also determined by the context the commission is operating in. For example, focusing on building the case for ETR (e.g. Danish Commission on Personal Income), building/exploring stakeholder support for ETR, development and evaluation of detailed proposals (de Waard Commission in the Netherlands) or implementation (eco-tax commission in Belgium).

Ensuring that the work of the commission is relevant to the needs of the political and public debate on ETR seems to be the lesson to draw from this. Commissions in countries where this has not been the case do not seem to be very successful (e.g. Ireland (first), France, Austria). For example, it is not useful, and may in fact be counterproductive, for a commission to develop detailed recommendations for ETR based on theoretical possibilities if the broader political actors do not consider them politically viable and the public are not aware or supportive of the case for ETR.

Functions can usefully be separated. The Dutch green tax commissions have focused on identifying and evaluating possible green tax options. The information from this process is then fed into other processes; in the case of the Van de Vaart Commission, the Dutch stakeholder NEPP process was used to develop and agree environmental policy, and in the case of the de Waard Commission as an input to the manifesto process. This can prevent measures from being disregarded prematurely.

With regard to communications, much of the work of the commissions seems to focus on the technical debate of measures amongst policy makers and/or experts. Less attention is given to the issues of public opinion and practical politics, how these might affect the development of ETR and the role that a commission may play in overcoming any problems in this area or opportunities it creates. It may be that work was done to address this point but that it is less well documented. Most of the commissions have stakeholder involvement to a greater or lesser degree but this only seems to be part of the broader communications effort required. The Irish Commission on Taxation and the UK Green Fiscal Commission are exceptions to this general trend. Both bodies utilised processes to raise awareness of their work with wider stakeholders and seek input to their work. In the case of the UK Green Fiscal Commission research was conducted to understand public attitudes to ETR including opinion polling and the holding of deliberative days to explore public attitudes to ETR and how views change as evidence is presented and the issues are debated (Green Fiscal Commission 2009).

Related to this point is that the activities of the commissions for the most part seem to be rather poorly documented. This will partly be a function of the age of some of the commissions and also the fact that they may be better documented in non-English sources which were not reviewed for this work. The Dutch commissions were an exception in this regard, and were better documented, reflecting a broader process of stakeholder and public engagement.

## Recommendations for future green tax commissions

It is clear from past Green Tax Commissions that their establishment, form and remit is very context-dependent, as illustrated by the comments of Schlegelmilch (2006, p. 2), then an official at the German Environmental Ministry, and active in Green Budget Germany, on the development of ETR in Germany:

Policy advice on ETR was particularly critical in the phase of gaining political support before its implementation, but also after its implementation when further improvements had to be made.... It is not easy to generalize the process and explore the lessons learned in order to help other countries which are also working on the implementation of economic policy instruments. Every country has a very special set of actors including the influence of certain people.... It is, however, important to really bring together different actors from the political parties, the bureaucracies, scientific institutions, industries to environmental NGOs with their counterparts in other countries to allow for creating a good network and for ensuring the same 'language' is spoken between the stakeholders.

However, bearing this in mind, it is possible to put forward the following issues to bear in mind in relation to the establishment and operations of future Green Tax Commissions:

1  *Stakeholder representation.* Not surprisingly, if the commission is to be effective and to get broad-based buy-in to its work it should include members from a full range of stakeholders which is representative of the broader political debate(s) on these issues. This should include representation from the main political parties to ensure the commission is seen as being politically impartial but also so it has a mechanism to become aware of the political and policy sensitivities around the developing debate and also feed into this. Other stakeholders including business, NGOs and academics will also need to be included. The membership should include expertise that covers not only issues associated with ETR but also broader expertise in tax and fiscal policy and issues of tax design.

2  *Independence.* Experience on whether a commission should be independent from government is mixed. Independence can make a commission better able to explore and debate sensitive issues. It may also be able to engage a broader range of political representation – an important consideration in building support for ETR – than a government-formed commission, even one set up at arm's length to government. Independence will also allow the commission to set its own agenda to a greater degree although it will have to work with the grain of external political and policy developments if it is to be successful. However, independent commissions can also struggle to gain influence with government once its outputs are produced.

3  *Relevance of outputs.* Any commission must carefully tailor its work and outputs so that they are relevant to the state of the debate on ETR in the country concerned and broader political context. In doing this it must identify and address the concerns that key decision makers and the public may have, and the state of public understanding of and information about ETR. It may need to produce a range of outputs at a general level on the various issues associated with ETR, e.g. the structure of ETR, impact on competitiveness and social equity, along with more detailed information on the likely impacts of particular options for ETR in the country concerned.

4   *Opening up not closing down the debate.* The focus of any commission should be on developing the possible options for ETR and understanding their implications, on opening up rather than narrowing down the debate on ETR. This approach mirrors the approach in the Netherlands where the commissions developed and evaluated a wide range of measures which were fed into other processes for development and decision as to whether they should be taken forward.

5   *Methods.* Important potential areas for work for a commission are:

- *Collation of evidence on the operation and implementation of green taxes.* Other commissions have predominantly focused on the technical aspects of the policy debate around the development of ETR. As with previous commissions, work in this area needs to involve the collation of existing evidence and generation of new insights through approaches including the modelling of a range of options for ETR, in order to understand the economic, environmental and social impacts of potential approaches to ETR.

- *Work to understand stakeholder and public attitudes to green taxes.* Any future commissions should conduct work that investigates broader stakeholder and public opinion on ETR as this will have a significant impact on any development of the policy, whatever the findings of the theoretical and modelling evidence may say as to its desirability.

- *Engagement with potential users of outputs.* Engaging at an early stage with policy makers and politicians who may be the user of commission outputs to understand their needs and concerns is likely to enhance the success of the commission's work.

- *Communication of work and findings of commissions.* Any commission needs to communicate and document its findings well. For the most part this does not seem to have been done very well by previous commissions. Many outputs are published as grey literature which is less well catalogued and harder to find in literature searches than journal articles or catalogued books (e.g. given a registered ISBN code). This is necessary for a number of reasons. Transparency on how findings and conclusions have been arrived at is important if the commission's outputs are to be trusted and tested by others. Second, the commission must make sure the evidence it assembles is distributed to appropriate target audiences if it is to be acted on or stimulate debate. Finally, the work will only be of value to future workers in this area if it is properly documented and accessible.

6   *Establish links with other existing commissions.* It obviously makes sense for any future commission to make contact and establish links with any existing commission elsewhere, in order to share work and exchange experience in this area. Currently this would comprise Green Budget Germany, Green Budget Europe and France who have commissions active and the Irish government who has an active policy debate on these issues.

**Annex**

*Table 9.a.1* Green tax commissions – summary details

| Country | Name | Date | Government formed | Membership | Objectives | Methods | Outputs | Recommendations | Impact |
|---------|------|------|-------------------|------------|------------|---------|---------|-----------------|--------|
| Austria | Austrian $CO_2$ Commission | 1993–1994 | Yes | Government and other stakeholders | Develop new systems of energy taxation based on ETR. | Research, modelling of impacts, strategy development and international workshops on ETR. | Research on baseline Austrian GHG emissions. Recommendations for meeting GHG targets including specific energy taxes. | A series of energy taxes; increased spending on ecological measures and cuts in labour taxes. | Only some measures implemented – gas and electricity taxes. |
| | Commission on Tax Reform | 1997–1999 | Yes | Government only | Explore environmental policy responses. | Not known. | Not known but may have been disbanded before reporting. | – | Limited – group disbanded after 1.5 years and prospect of ETR ruled out by government. |
| Belgium | Follow-up Commission Eco-taxes | 1993–2000 | Yes | 13 experts | To review implementation of Belgian eco-taxes and develop new proposals. | Review process – no details available. | As recommendations. | Changes in design and rates of existing ecotaxes and proposals for new ones. | Significant – but institutional position within government contributed to this. Improved implementation of eco-taxes. |
| Denmark | Commission on Personal Income | Reported in 1992 | Yes | Not known | To reduce taxes on personal income and fund resulting fiscal gap through green taxes amongst other measures. | Included economic modelling of impacts of interaction between fiscal revenues, employment, environment and competitiveness. | Report to government. | Reduced tax revenues from income taxes, increased revenues from social security contributions, broadening of tax base and environmental taxes. | Significant – informed programme of measures implemented from 1993 onwards. |
| | Inter-ministerial committee on business energy taxation | 1993–1995 | Yes | Government inter-ministerial committee | Development of business energy taxes with particular regard to impact on international competitiveness. | Not known. | Recommendations on business energy taxes. | Ongoing programmes of measures including $CO_2$ and $SO_2$ taxes and recycling mechanisms. | Recommendations implemented and on review found to be economically efficient and acceptable in regard to competitiveness. |

*continued*

Table 9.a.1 Continued

| Country | Name | Date | Government formed | Membership | Objectives | Methods | Outputs | Recommendations | Impact |
|---|---|---|---|---|---|---|---|---|---|
| France | Taxe Générale sur les Activités les Polluantes | 1999–2000 | Yes | – | To unify and simplify several existing environmental fees/taxes and to implement new ones | – | – | | Ruled unconstitutional in 2000 and political conflicts ended any chance of reform. |
| | Outils Économiques et Développement Durable | February 2006 to July 2007 | Yes | 35 members from government, MPs, academia, business, NGOs. Gov. secretariat. Chair Jean-Pierre Landau (Banque de France). | To develop proposals for economic instruments to address climate change, loss of biodiversity and traffic congestion. | Expert and stakeholder working group with support of secretariat. | Les Instruments Economiques Du Developpement Durable. Rapport du Groupe de Travail présidé par Jean-Pierre LANDAU July 2007 | Proposed introduction of price signal for GHG emissions, congestion charging and HGV kms-based tax, local tax instruments for biodiversity. Consistency of implementation critical | Process of government consideration of carbon tax stalled due to record oil price |
| | Rocard Commission (Source: Senit 2012) | July to August 2009 | Yes | Members from government, employers and labour representatives, economists. Chaired by former Prime Minister Michel Rocard. | Practical design of tax – tax base and rate and mechanism to recycle revenues. | Expert and stakeholder group. | Report to government | Mixed carbon/energy tax base progressively increasing from €32/t $CO_2$ to €100/t $CO_2$ in 2030. | Proposal developed by government into legislation but challenge following implementation invalidated tax and carbon tax project postponed following major defeat of governing party regional elections |
| | Comité pour la fiscalité écologique | December 2012 to present | Yes | 40 members including elected officials, employees, representatives of employers, local authorities, associations and NGOs. Chaired by Prof. Christian de Perthuis. | To develop proposals finance bill 2014 and in longer term to consider the development of environmental taxation in the context of an overall reform of tax and social security deductions | Expert and stakeholder group. | Initial report to government March 2013 | Reduce tax advantage of diesel over petrol. | Too early to say. |

| Country | Name | | Timeframe | Membership | Focus | Methods | Outputs | Reports | Outcome |
|---|---|---|---|---|---|---|---|---|---|
| Germany | Green Budget Germany | No | 1994 to present | 3 staff, 5 member executive board, 22 member advisory board | Its focus to include support to development of market-oriented eco-fiscal policy: environmental taxation, emission trading, removal of environmentally harmful subsidies, promotion of renewable energies and green growth. | Operates as a pressure group – events, briefings, media work and other networking/lobbying activities | As methods. | Wide ranging reports on ETR. | Appears to have effectively contributed to development of ETR debate. |
| Ireland | Environmental Tax Group | Yes | 1996–1997 and possibly on to 2002 | Government only | To examine strategic impact of taxation on environmental policy and to develop measures with particular reference to competitiveness. | Not known. | Reports to Tax Strategy Group of Irish Deptartment of Finance | In 1999 proposals for diesel tax rebates, vehicle registration taxes, VAT on electricity, plastic bag tax. Low sulphur diesel, framework on GHG taxes. In 2002 an excise-type tax was proposed on all fossil fuels across all sectors, directly related to the amount of CO$_2$ emissions. | Low, other than plastic bag tax. Work led to consultation on carbon tax in 2003 and subsequent abandonment of carbon tax proposal in 2004. |
| Ireland | Commission on Taxation | Yes | February 2008 to July 2009 | 17 members drawn from range of expert and stakeholder backgrounds. | 'To review the structure, efficiency and appropriateness of the Irish taxation system' including 'Investigate fiscal measures to protect and enhance the environment including the introduction of a carbon tax'. | Meetings, subgroups, plenary sessions; submissions; oral hearings and other consultations. | Commission on taxation Report 2009 (561 pages) | Comprehensive recommendations covering all aspects of tax policy and implementation. Recommendations in regard to environment: carbon tax, metered water charges, waste charges, road charges, not taking gains from EU ETS, allowances, accelerated capital allowances for energy efficient equipment, favourable VAT rates for environmental goods and the offset of R&D tax credits against employer PRSI costs and broad acknowledge of importance of green economy. | Carbon tax implemented as recommended. |

*continued*

Table 9.a.1 Continued

| Country | Name | Date | Government formed | Membership | Objectives | Methods | Outputs | Recommendations | Impact |
|---|---|---|---|---|---|---|---|---|---|
| The Netherlands | Van de Vaart Commission | 1995–1997 | Yes | Chaired by government (Ministry of Finance) with environment, fiscal, employer reps and ex-MPs. | Evaluation of tax system as a whole to see if it could work better for environment. | Rigorous evaluation of possible environmental taxes. | Three reports fed into Dutch NEPP process. | First two reports on greening taxes on the purchase, ownership and use of private cars. Third report on long-term options for greening the tax system. | By 1998, 14 per cent of total tax revenue was raised by green taxes with many of the recommendations from the first and second reports having been implemented. Remaining proposals included in White Paper on further tax reform including environmental taxes. |
| | De Waard Commission | 2000–2001 | Yes | Chaired by government (Ministry of Finance) with environment, fiscal, employer reps and ex-MPs. | To evaluate existing measures, the anticipated effects of new measures and define a new agenda for next term of government. | Rigorous evaluation of possible environmental taxes | Report fed into manifesto process. | Evaluation of large number of possible taxes | Low – but effectively fulfilled objective of scoping future agenda and adoption of many of previous commissions' recommendations and left more challenging options to be taken forward. |
| Norway | Norwegian Governmental Commission on Green Taxes | 1994–1996 (and 1990–1992) | Yes | 116 members from wide range of backgrounds. | To investigate how to shift tax system from labour to environment in a long term context. | Assessment of impact of ETR including economic modelling. | Report to Ministry of Finance including recommendations and estimates of revenue and uses it could be put to. | Guidelines for changing basic framework conditions of economy and specific tax proposals for short-, medium- and long-term measures. | Government followed up recommendations with White Paper on Climate Change but partly voted down but notably an aviation tax was implemented. |

| | | | | | | | | | |
|---|---|---|---|---|---|---|---|---|---|
| Sweden | Commission on Green Taxation | 1995–1996 | Yes | Parliamentary Commission of MPs and with expert input. | Assessment of ETR's impact on environment, employment, resource use and tax revenues and stability of tax revenues. | Included economic modelling of options and assessment of efficiency of environmental taxes. | Report in 1997. | General principles to steer development of ETR rather than specific measures. | Contribution to development of taxes on energy, carbon and sulphur. |
| UK | Green Fiscal Commission | May 2007 to October 2009 | No, independent | 22 members with representative range of social, economic and political stakeholders. | To prepare the ground for a significant programme of green fiscal reform in the UK, in terms both of assembling the evidence base for such a reform, and of raising stakeholder and public awareness of it. | Commission meetings, evidence review, collation and synthesis, generation of new evidence including economic modelling and research to investigate public opinion, stakeholder engagement; communication and dissemination activity (briefings, reports, media work conference) | Final report in October 2009 and series of briefings and research reports on main areas of work. | Recommendations not made but evidence presented on key messages on green fiscal reform: higher green taxes, implemented as a tax shift, could by themselves enable the UK to meet its statutory carbon emission reduction targets. The tax would also provide a boost to the development of low carbon industries and, in doing so, provide additional employment. | Significant contribution to evidence base and political debate on ETR in UK which led to commitments in election manifesto of party leading coalition government following 2010 election and in Coalition Agreement of the government formed. |

## Notes

1 The overall tax ratio comprises the sum total of all taxes and social security contributions as a percentage of gross domestic product.
2 The official literature for the Commission formed in 2000 refers to itself as the 'Second Dutch Green Tax Commission' although the existence of the 1990 commission would suggest it is the third. To avoid confusion we refer to the two later commissions by the names of their chairs.
3 The recommendations of each of the three Van der Vaart Commission reports are summarised in Vemeend and Van der Vaart (1998, pp. 106, 108).

## References

Batho, D. (2012) 'Installation du Comité pour la fiscalité écologique'. Ministerial speech given on 18 December. Online at www.developpement-durable.gouv.fr/IMG/pdf/_Intrduction_fiscalite_ecologique_DB_ok_.pdf.

Brännlund, R. and Gren, I. (1999) *Green Taxes: Economic Theory and Empirical Evidence from Scandinavia.* Edward Elgar, Cheltenham.

Brännlund, R. and Kristom, B. (1997) 'Energy and Environmental Taxation in Sweden: Some Experiences from the Swedish Green Tax Commission'. Invited Paper 'Environmental Implications of Market Based Policy Instruments', Gothenburg, 20–21 November.

Bretteville, C. and Søfting, G.B. (2000) 'From Taxes to Permits? The Norwegian Climate Policy Debate'. CICERO Report 2000:6. CICERO, Oslo.

Commission on Taxation (2009) 'Commission on Taxation Report 2009'. The Stationery Office, Dublin.

Convery, F. (2012) 'Budget 2013L Three Cheers for the Carbon Tax'. Online at www.publicpolicy.ie/budget-2013-three-cheers-for-the-carbon-tax/.

Convery, F., McDonnell, S. and Ferreira, S. (2005) 'The Most Popular Tax in Europe? Lessons from the Irish Plastic Bag Levy'. Paper submitted to the EAERE conference, Bremen, 23–26 June 2005. Cited in OECD (2006) *The Political Economy of Environmentally Related Taxes*, OECD, Paris, p. 124.

Danish Energy Agency (2000) 'Green Taxes for Trade and Industry: Description and Evaluation'. Danish Energy Agency, Copenhagen.

Department of Finance (1999) 'Environmental Tax Policy: Interim Report by Chairman of Working Party'. Tax Strategy Group Paper TSG 99/29, Department of Finance, Dublin.

Department of Finance (2002) 'National Climate Strategy: Proposal for a Carbon Energy Tax'. Tax Strategy Group Paper TSG/02/23, Department of Finance, Dublin.

Department of Finance (2004a) 'Carbon Taxation Consultation Process: Report on Submissions Received'. Department of Finance, Dublin.

Department of Finance (2004b) 'McCreevy Abandons Carbon Tax'. Press Release 10 September. Online at www.finance.gov.ie/viewdoc.asp?DocID=2608&CatID=1&StartDate=01+January+2004&m=.

Department of Finance (2009) 'Financial Statement of the Minister for Finance Mr Brian Lenihan,T.D'. Department of Finance, Dublin, 9 December.

Department of Finance (2012) 'Financial Statement of the Minister for Finance Mr. Michael Noonan, T.D'. Department of Finance, Dublin, 5 December.

Deroubaix, J.F. (2006) 'The Rise and Fall of French Ecological Tax Reform: Social Acceptability Versus Political Feasibility in the Energy Tax Implementation Process', *Energy Policy* 34(8), 940–949.

Dutch Green Tax Commission (2001) ' "Greening" the Tax System: An Exploration of Ways to Alleviate Environmental Pressure by Fiscal Means'. Summary translation of the official report of the Dutch Green Tax Commission chaired by Jan de Waard. No publisher details or location given. Online at www.who.int/ifcs/documents/standing-committee/greentax_nl.pdf.

Ekins, P. and Speck, S. (2000) 'Proposals of Environmental Fiscal Reforms and the Obstacles to their Implementation', *Journal of Environmental Policy and Planning* 2(2), 93–114.

Ekins, P., Kleinman, H., Bell, S. and Venn, A. (2010) 'Two Unannounced Environmental Tax Reforms in the UK: The Fuel Duty Escalator and Income Tax in the 1990s', *Ecological Economics* 69(7), 1561–1568.

Ekins, P., Summerton, P., Thoung, C. and Lee, D (2012) 'A Major Environmental Tax Reform for the UK: Results for the Economy, Employment and the Environment', *Environmental and Resource Economics* 50(3), 447–474.

*ENDS Europe* (2000) 'Sweden Swims Against the Tide Over Fuel Tax', 21 September. Online at www.endseurope.com/999/sweden-swims-against-the-tide-over-fuel-tax.

*ENDS Europe* (2006a) 'Call to Restart German Energy Tax Escalator', 8 May. Online at www.endseurope.com/index.cfm?go=11857.

*ENDS Europe* (2006b) 'Sweden Abandons its Green Tax Dream', 17 October. Online at www.endseurope.com/12548/sweden-abandons-its-green-tax-shift-dream.

*ENDS Europe* (2013) 'French Minister backs Changes to Fuel Taxation', 5 March. Online at www.endseurope.com/30894/french-minister-backs-changes-to-fuel-taxation.

Green Budget Germany (2013) 'About Us'. Pages on Green Budget Germany website. Online at www.foes.de/ueber-uns/.

Green Fiscal Commission (2009) 'The Case for Green Fiscal Reform: The Final Report of the Green Fiscal Commission'. Policy Studies Institute, London.

HMG (Her Majesty's Government) (2010) 'The Coalition: Our Programme for Government'. HMG, London, May. www.gov.uk/government/uploads/system/uploads/attachment_data/file/78977/coalition_programme_for_government.pdf.

Hoerner, J.A. and Bosquet, B. (2001) 'Environmental Tax Reform: The European Experience'. Centre for a Sustainable Economy, Washington DC. Online at http://rprogress.org/publications/2001/eurosurvey_2001.pdf.

Ichay, F. (n.d.) 'Creation of a Committee for the Environmental Tax System'. Briefing by Ichay & Mullenex published 13 February 2013. Online at www.ichay-mullenex.com/download/creation-of-a-committee-for-the-environmental-tax-system.pdf.

Klok, J., Larsen, A., Dahl, A. and Hansen, K. (2006) 'Ecological Tax Reform in Denmark: History and Social Acceptability', *Energy Policy* 34(8), 905–916.

Landau. J.-P. (2007) 'Les Instruments Economiques du Developpement Durable Rapport du Groupe de Travail présidé par Jean-Pierre LANDAU'. Online at www.minefe.gouv.fr/services/som_rapports.php.

Martin, C. (1997) 'Eco-taxes in the EU'. Online at www.chemsoc.org/chembytes/ezine/1997/eco.htm.

Mortensen, J.B. and Hauch, J. (1999) 'Government Commissions on Green Taxes in Denmark', in Brännlund, R. and Gren, I. (eds) *Green Taxes: Economic Theory and Empirical Evidence from Scandinavia*, Edward Elgar, Cheltenham.

NGTC (Norwegian Green Tax Commission) (1996) 'Policies for a better environment and high employment'. An English summary of the Norwegian Green Tax Commission, NGTC, Oslo.

OECD (2001) 'Encouraging Environmentally Sustainable Growth in Belgium. Economic Department Working Paper 300', OECD, Paris. Online at www.oecd.org/belgium/1892052.pdf.

OECD (2006) *The Political Economy of Environmentally Related Taxes*, OECD, Paris

Schlegelmilch, K. (2006) 'The Experiences with Green Budget Reform in the EU and especially in Germany'. Unpublished paper based on Schlegelmilch, K. (2005) 'Experiences with the Ecological Tax Reform in Germany History, Design, Experiences, Impacts and Marketing', in Böhringer, C. and Lange, A. (eds) *Applied Research in Environmental Economics*, ZEW Economic Studies Vol. 31, Physica-Verlag, Heidelberg.

Schlegelmilch, K. (1997) 'Green Tax Commission, Environmental Policy Research Brief, Number 4'. Environmental Institute, University College, Dublin. Online at www.ucd.ie/t4cms/pb-ep-04.pdf.

Scottish Executive (2005) 'Proposed Plastic Bag Levy: Extended Impact Assessment Final Report Volume 1: Main Report'. Environment Group Research Report 2005/06. Online at www.scotland.gov.uk/Resource/Doc/57346/0016899.pdf.

Senit, C.-A. (2012) 'The Politics of Carbon Taxation in France: Preferences, Institutions and Ideologies'. Working Paper No. 20, IDDRI, Paris, France. Online at www.iddri.org/Publications/Collections/Idees-pour-le-debat/WP2012_CAS_politics%20of%20taxation%20france.pdf.

Shaw, B. and Venn, A. (2007) 'A Green Fiscal Commission for the UK: Scoping Research: Review of the processes and recommendations of Green Tax Commissions in Other Countries and Recommendations for the Operation of the UK GFC', mimeo, Policy Studies Institute, London.

Sustainable Development Commission (2006) 'Green Tax Commissions'. Unpublished draft report, Sustainable Development Commission, London.

Swedish Environmental Protection Agency (2004) 'Developing Green Taxation'. A Summary of a Government Assignment Report 5390, Swedish Environmental Protection Agency, Stockholm.

UNFCCC (1995) 'Executive Summary of the National Communication of Austria Submitted under Articles 4 and 12 of the United Nations Framework Convention on Climate Change'. Online at http://unfccc.int/resource/docs/nc/aut01.htm.

Vermeend, W. and Van der Vaart, J. (1998) *Greening Taxes: The Dutch Model*, Kluwer Law International, London.

Wurzel, K.W., Brückner, L., Jordan, A. and Zito, A.R. (2003) 'Struggling to Leave Behind a Highly Regulatory Past? "New" Environmental Policy Instruments in Austria', *Environmental Politics* 12(1), 51–72.

# Part III

# Carbon-energy taxation in East Asia

Features and challenges

# 10 Energy tax reform toward a sustainable low carbon economy

## The case of Japan

*Soocheol Lee and Kazuhiro Ueta*

## Introduction

Japan currently has three types of energy-related taxes (referred to generally as "energy taxes" in this chapter): (1) taxes that are levied when energy sources are imported or extracted, such as the Petroleum and Coal Tax, (2) taxes that are levied on the use of transportation fuels, such as the Gasoline Tax, and (3) the tax levied on the energy conversion sector, namely, the Tax for Promoting the Development of Power Plants.[1]

These energy taxes are not directly intended to control various negative externalities resulting from energy uses, such as carbon dioxide emissions. While it is true that these taxes may have contributed to reductions in the burden on the environment and can thus be regarded as second-best environmental taxes, Japan's energy taxes had been special-purpose revenue sources earmarked to be allocated to designated uses, including subsidies for the coal and petroleum industry, the construction and maintenance of roads, and the development of thermal power plants centering on fossil-fuel or nuclear power generation. In other words, it can be said that, in terms of the usage of tax revenues, Japan's energy taxes have rather contributed to increases in external diseconomy by stimulating energy consumption and destructing environment.

Various reforms on energy taxes have been undertaken in recent years. For example, the Japanese government introduced carbon tax, which is a novelty for Asian countries, at the tax rate of 289 yen/$CO_2$ ton, under the name of "anti-global warming tax" in October 2012. Yet, because of the very low tax rate, there are doubts whether this tax will provide sufficient incentives to corporations and people to reduce carbon dioxide emissions. Although traditional special accounts that received energy-related, special-purpose tax revenues have been abolished or combined, similar special accounts, under the name "The Special Account for Energy Measurement", are newly created at the same time, indicating that the traditional basic framework, in which revenues from energy taxes have been treated as special-purpose revenue sources like above, has not changed.

In this context, this chapter clarifies the theoretical foundations for and characteristics of energy taxes, and provides a detailed examination of the effects of

Japan's energy taxes on carbon dioxide emissions, focusing on how they are levied and used. This chapter also discusses how the system of energy taxes should be reformed in order to promote energy sustainability and the transition to a sustainable low carbon economy.

## The theoretical foundations for energy taxation

The theoretical foundations for energy taxation are explained as follows. Hotelling (1931) offers a first theoretical basis for energy taxation with an analysis of the optimal rule for the usage of petroleum and other resources that takes into account their characteristics as exhaustible resources. He explains a long-term upward tendency of energy prices by deriving what has become known as Hotelling's rule in which the net price of an exhaustible resource rises every period at the same rate as the interest rate under the efficient extraction path.[2] Hotelling also points out that exhaustible resources tend to be extracted and consumed excessively at low prices because of producers' short-term profit-seeking motive, and that resource extractions and logging need to be controlled for the purpose of resource conservation and for future generations. As a way to achieve this, he suggests introducing unit-based taxes on the extraction of exhaustible resources.[3]

Hartwick (1977) and Dasgupta and Mitra (1983) provide another theoretical basis for energy taxation. They further develop Hotelling's theory and point out the problem of intergenerational equity in using exhaustible resources. For example, Hartwick derives a result, which has become referred to as Hartwick's rule, that when the rent generated by extracting exhaustible resources is used to accumulate reproducible capital, sustainable consumption is possible. Under Hartwick's rule, a decrease in exhaustible resources means reduced productive capacity. Therefore, it is implied that such reduced productive capacity must be balanced with an accumulation of manufactured capital (such as machinery and equipment) that is substitutable for exhaustible resources in order to maintain the same consumption level between the present and future generations.

Swerling (1962) and Takahashi (1972) consider resource and energy issues as an adjustment problem between the speed of exhausting natural resources in resource-abundant countries and the speed of developing alternative energies and making industry adjustments in resource-consuming countries. For the adjustment of the two speeds, they propose the introduction of international taxation on exhaustible resources and the use of the tax revenues for the development of alternative energy sources such as nuclear fusion, and solar energy.

A third basis for energy taxation is the internalization of external diseconomy resulting from the use of energy. This is different from the first and second theoretical bases in that it may not provide an essential justification for energy taxes. This third basis for energy taxation emerged from the demand in a period in the past for curbing rapid increases in external costs associated with energy uses, and its theoretical foundation dates back to Pigou (1920). The tax proposed by

Pigou is aimed to internalize of external costs resulting from the consumption of the energy source. This tax known as the Pigouvian tax has become the starting point of today's environmental taxes.

In sum, there are three theoretical bases for energy taxation: (1) the realization of intergenerational equity through the optimal use of exhaustible resources, (2) the securing of revenue sources for the accumulation of capital substitutable for exhaustible energy sources, and (3) the internalization of external diseconomy resulting from energy consumption.

Based on the discussions in this section, this chapter takes the position that a tax system which satisfies simultaneously these three bases is the ideal energy tax system. In other words, the energy tax system can play a crucial role in enhancing a sustainable low carbon economy.

## The current state and characteristics of Japan's energy taxes

As mentioned earlier, Japan's energy taxes are divided into three main categories: (1) the Petroleum and Coal Tax, (2) the transportation fuel taxes, and (3) the Tax for Promoting the Development of Power Plants.

The revenues from these energy taxes amount to 4.822 billion yen based on the FY2009 budgetary standards. Among them, the revenue from transportation fuel taxes levied on products like gasoline and light fuel oil is 3.961 billion yen which is 82.1 percent of the total energy tax revenues. In contrast to largely declining national tax revenues due to the long-term economic stagnation since the 1990s, revenues from energy taxes have constantly continued to rise. Therefore, the proportion of the energy tax revenues to the entire national tax revenue has been higher than 10 percent since FY2000, as opposed to 6.7 percent for FY1990. They thus constitute a major source of tax revenues (see Table 10.1).[4]

The Tariffs on Crude Oil and Petroleum Products had been a source of subsidies mainly supporting the domestic coal industry since their introduction in 1955. The subsidy to the coal industry was based on industrial policy considerations such as the maintenance of employment and the protection of the domestic coal industry as a supplier of alternative energy sources to petroleum.[5] However, for FY2007 and subsequent years, it was abolished along with the termination of a coal policy.

The Petroleum and Coal Tax was introduced in 1978 for the purpose of obtaining a revenue source mainly to secure stable supply of petroleum. Similarly to the Tariffs on Crude Oil and Petroleum Products, it is levied on crude oil (as well as imported petroleum products). Therefore, basically the same tax rates are uniformly applied in upstream sectors regardless of whether the oil is used for electricity generation, industrial purposes, transportation, or household use. The tax rates are 2.04 yen/l for crude oil and imported petroleum products, 1.08 yen/kg for imported liquefied petroleum gas (LPG), and 0.7 yen/kg for coal, and the tax revenues amount to 510 billion yen (in the FY2009 budget).

The original purpose of the Petroleum and Coal Tax derives from special conditions surrounding petroleum energy such as ensuring energy security.

Table 10.1 Revenues from the energy taxes (billion yen)

| Category | Fiscal year (FY) | | | | | | | | |
|---|---|---|---|---|---|---|---|---|---|
| | 1970 | 1975 | 1980 | 1985 | 1990 | 1995 | 2000 | 2005 | 2009 |
| Total energy tax revenues | 892 | 1363 | 3024 | 3381 | 4193 | 5124 | 5323 | 5335 | 4822 |
| Tariffs on Crude Oil and Petroleum Products | 134 | 145 | 151 | 131 | 103 | 82 | 55 | 45 | – |
| Petroleum and Coal Tax | – | – | 404 | 400 | 487 | 513 | 489 | 493 | 510 |
| Total transportation fuel taxes | 758 | 1217 | 2360 | 2616 | 3308 | 4190 | 4404 | 4439 | 3961 |
| Petroleum Gas Tax | 24 | 28 | 30 | 31 | 31 | 31 | 28 | 28 | 26 |
| Gasoline Tax | 589 | 974 | 1826 | 1968 | 2367 | 2726 | 3065 | 3220 | 2909 |
| (National) | 499 | 824 | 1547 | 1668 | 2007 | 2463 | 2769 | 2908 | 2628 |
| (Local) | 90 | 150 | 278 | 300 | 361 | 263 | 296 | 311 | 281 |
| Light Fuel Oil Tax | 144 | 194 | 447 | 556 | 833 | 1332 | 1207 | 1086 | 928 |
| Aircraft Fuel Tax | – | 22 | 58 | 62 | 76 | 101 | 104 | 105 | 98 |
| Tax for Promoting the Development of Power Plants | – | – | 108 | 233 | 295 | 339 | 375 | 359 | 351 |
| Total national tax revenue | 7775 | 10,401 | 28,373 | 39,150 | 62,780 | 54,963 | 52,721 | 52,290 | 47,815 |

Source: Ministry of Finance (1970–2009).

Notes
1 Among energy-related taxes, the Light Fuel Oil Tax is a local tax.
2 The tariff on crude oil was abolished for FY2007 and the subsequent years.

Table 10.2 Rates of major energy taxes

| Tax category | Gasoline Tax (yen/l) | Petroleum and Coal Tax | | | TPDE (yen/kWh) |
|---|---|---|---|---|---|
| | | Crude oil and imported petroleum products (yen/l) | Natural gas (yen/kg) | Coal (yen/kg) | |
| October 1, 1980 | 53.8 | 3.5 | – | – | 0.300 |
| October 1, 1983 | 53.8 | 3.5 | – | – | 0.445 |
| September 1, 1984 | 53.8 | 4.7 | 1.2 | – | 0.445 |
| August 1, 1988 | 53.8 | 2.04 | 0.72 | – | 0.445 |
| October 1, 2003 | 53.8 | 2.04 | 0.84 | 0.23 | 0.425 |
| April 1, 2005 | 53.8 | 2.04 | 0.96 | 0.46 | 0.400 |
| April 1, 2007 | 53.8 | 2.04 | 1.08 | 0.70 | 0.375 |
| May 1, 2008 | 53.8 | 2.04 | 1.08 | 0.70 | 0.375 |
| April 1, 2012 | 53.8 | 2.04 | 1.08 | 0.70 | 0.375 |

Source: Petroleum Association of Japan (1980–2012).

Note
TPDE is abbreviation form of Tax for Promoting the Development of Power Plants.

Therefore, the purpose of introducing the tax itself had no relation with energy sustainability or countermeasures to global warming. However, it should be noted that since the 1980s, part of the tax revenue has been allocated to the promotion of energy conservation and new energies.[6]

The taxes related to transportation fuel, such as the Gasoline Tax and the Light Fuel Oil Transaction Tax, have been levied based on the beneficiary pays principle that those who use the roads and airports should pay a usage fee for such infrastructure. Actually, all of the revenues from these taxes have long been allocated to the construction and improvement of roads and airports as special-purpose revenues.[7] These taxes were introduced to secure stable revenues that would cover the construction cost of transportation infrastructure which had been lagging behind during the period of post-war economic recovery. Since then, they have played a significant role in the construction and improvement of the roads and airports in Japan. In other words, the taxes related to transportation fuel can be regarded as the driving force of the creation of a fiscal structure prioritizing road improvement and have promoted an increase in the share of public projects in the government budget. Also, road constructions financed by these taxes stimulated and spread the use of automobiles, which contributed to not only increase in transportation fuel consumption but also rising revenues from these taxes.

Contrasting with the aforementioned energy taxes directly levied on the use of fossil fuel, the Tax for Promoting the Development of Thermal power plants is levied when electric power converted from fossil fuel is sold to entities on the demand side. In order to encourage the construction of power plants centering on nuclear power generation but including fossil-fuel and hydropower generation, the tax was created in 1974 as a earmarked tax that would generate revenues under the objective of implementing measures related to the Power Plants establishment, such as improving public facilities in their neighborhood and conducting projects to develop local economies. A total of 40 to 60 percent of the tax revenue has been used to ensure energy security through, for example, measures related to the establishment of nuclear power generation facilities.

## Energy taxes revenues and their uses

This section examines in detail the use of energy tax revenues by dividing them into three major categories: (1) revenues from upstream taxes such as the Petroleum and Coal Tax, (2) revenues from downstream taxes such as the taxes related to transportation fuel, and (3) revenues from the Tax for Promoting the Development of Power Plants.

### *Petroleum and Coal Taxes revenues and their uses*

The Petroleum Tax was created in 1979, which induced the establishment of the Special Account for Measures Concerning Coal, Petroleum, and

Petroleum-Substituting Energies in 1980. The Petroleum Tax was modified into the Petroleum and Coal Tax in 2003. The revenue from the Tariff on Crude Oil and Petroleum Products was used mainly for measures to support the coal industry.[8] The purpose of the Petroleum and Coal Tax was changed into predominantly enhancing petroleum security and petroleum-substituting energies. In 1993, the special account was revised and became the Special Account for Improving the Supply and Demand Structure for Coal, Petroleum, and Energy, and its use for energy conservation programs and the development of new energies became emphasized. In 2002, as the support for the coal industry based on the revenue from the Tariff on Crude Oil and Petroleum Products was terminated, the special account was renamed the "Special Account for Improving the Supply and Demand Structure for Petroleum and Energy" (hereafter, "Petroleum Special Account").

In addition, based on Article 36 of the Act on Promoting Administrative Reforms to Realize Simplified and Efficient Government (a.k.a. the Administrative Reform Promotion Act) which took effect in June 2006, the Petroleum Special Account and the Special Account for Promoting the Development of Power Plants (hereafter, "Electricity Special Account"), whose main revenue source was the Tax for Promoting the Development of Power Plants, were integrated into the Special Account for Energy Measures, starting in FY2007 (Figure 10.1).

The Special Account for Energy Measures is a special account created to clarify the accounting of funds used for stabilization of fuel supply, structural improvement of energy supply and demand, power plant locations, and diversification of electric power generation. It is divided into the Subaccount for Energy Supply and Demand and the Subaccount for Promoting the Development of Power Plants. The nature of this special account shows that, in effect, the former and latter subaccounts succeed the abolished Petroleum Special Account and Electricity Special Account.

Until FY2000, 20 to 30 percent of the Petroleum Special Account was used for providing assistance for the coal industry, and 60 to 70 percent of the Account was used for petroleum-related measures such as oilfield development and efficiency improvement in oil production and transportation, which shows the inclination of the special account toward supporting the existing energy industry. However, in addition to such support, the share of funds used for new energy source promotion, carbon dioxide reduction projects, and energy conservation, began to rise rapidly in most recent years in the Special Account for Energy Measures (the Subaccount for Energy Supply and Demand), which succeeded the Petroleum Special Account.

### Transportation fuel taxes revenues and their uses

Prior to the termination of the Special Account for Road Improvement (FY2008 and earlier years), all revenues from the taxes related to transportation fuel were used as special-purpose revenues for road construction and improvement along

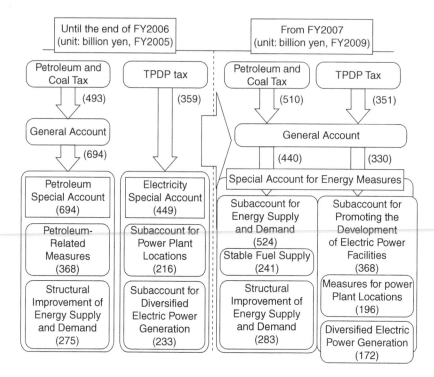

*Figure 10.1* Revised accounting system for the revenues from the Petroleum and Coal Tax and the Tax for Promoting the Development of Power Plants (billion yen) (source: created by the authors based on documents from the Ministry of Economy, Trade and Industry).

Notes
1  Due to accounting surplus, etc., tax revenues and values in the accounts do not match.
2  TPDP is abbreviated form of Tax for Promoting the Development of Power Plants

with automobile-related taxes such as the Automobile Weight Tax (715 billion yen in 2008 fiscal year) and the Automobile Acquisition Tax (485 billion yen in 2008 fiscal year). For the taxes related to transportation fuel and automobile-related taxes, the relationship between benefits and burden is relatively clear, as mentioned earlier. Therefore, taxpayers have accepted them with little resistance.

The share of the taxes related to transportation fuel in the revenue sources for road construction and improvement is 46.1 percent (FY2005 budget). The automobile-related taxes (for central government plus the local governments), which are the combination of this and the "automobile body taxes" that include the Automobile Acquisition Tax and the Automobile Weight Tax, amount to 61.4 percent.

In recent years, environmental issues faced by roadside areas have become a social problem, and budgets for addressing such issues have been created since

FY1998. These budgets are allocated to, for example, projects for improving the roadside environment by implementing measures such as the use of low-noise pavement for roadside areas where the level of noise or air pollution exceeds environmental standards. The budgets have been used recently for infrastructure improvement for monorail and streetcar systems, improvement for optical fiber networks along major roads, a project to place electric power cables in underground common utility ducts, comprehensive support for the vitalization of towns, maintenance of roadside trees, assistance for the purchase of equipment to remove particulates from diesel engines, assistance for creating electronic toll collection lanes, and the construction of underground shopping arcades.

In December 2005, the central government issued the Basic Guideline on the Reassessment of the Revenue Sources Designated for Road Projects, and the system governing revenue sources designated for road projects was abolished on April 30, 2009. The taxes related to transportation fuel, which had provided revenues to the Special Account for Road Improvement and the Special Account for Airport Improvement as special-purpose revenue sources, became general revenue sources. The two abolished special accounts were succeeded by the Subaccount for Road Improvement and the Subaccount for Airport Improvement of the Special Account for Infrastructure Development which began to receive necessary funds for road and airport projects from the general account (Table 10.3). The FY2009 revenue budgets for these subaccounts were 2483 billion yen and 530 billion yen, respectively. Among them, funds from the general account were 1639 billion yen and 142 billion yen, respectively. After all, the traditional basic framework, in which revenues from automobile-related taxes have been treated as special-purpose revenue sources, has not changed after the elimination of the special-purpose revenue sources.

### Tax Revenue from the Tax for Promoting the Development of Power Plants and its use

The revenue from the Tax for Promoting the Development of Power Plants entered the Special Account for Promoting the Development of Power Plants (Electricity Special Account) and has contributed to the expansion of the capacity to supply

*Table 10.3* Breakdown of the special account for infrastructure development (billion yen)

|                                    | 2008 | 2009 |
|------------------------------------|------|------|
| Subaccount for River Improvement   | 1159 | 1047 |
| Subaccount for Road Improvement    | 3555 | 2367 |
| Subaccount for Harbor Projects     | 352  | 303  |
| Subaccount for Airport Improvement | 556  | 530  |
| Subaccount for Operations          | 328  | 303  |
| Total                              | 5944 | 4550 |

Source: compiled by the authors based on data from Zaisei Chosakai (2008–2009).

electric power by promoting the development of power plants centering on nuclear power generation. With this revenue, the structure of electric power supply, which relied heavily on petroleum, has shifted toward nuclear power. In the FY2005 budgetary standards, the share of the budgets related to the development of nuclear power generation facilities in the Electricity Special Account reached approximately 76 percent.

It is notable that funds from the Subaccount for Power Plant Locations of the Electricity Special Account have been used to raise people's acceptance to nuclear facilities, being designated as subsidies to promote economic and social development in regions where nuclear power plants have been located. As a result, potential risks of radioactive pollution caused by nuclear power plants or risks associated with radioactive waste disposal have increased (Shimizu 2002). Also, the Subaccount for Diversified Electric Power Generation, which is intended for a shift away from petroleum-based power generation, has been used to develop a nuclear fuel cycle or construct hydro-power generation facilities, increasing potential risks and social costs.

In addition to the development and promotion of nuclear power generation and hydro-power generation, the development and promotion of new energies, such as solar power generation, power generation from waste, wind power generation, and geothermal power generation, have been included since the second half of the 1980s. The share of such development and promotion of new energies in the revenue from the Tax for Promoting the Development of Power Plants increased from 13.3 percent in FY1990 to 18.3 percent in FY2000, but decreased to 10.2 percent in 2005.

The Electricity Special Account and the Petroleum Special Account contribute to the development of petroleum-substituting energies at the level of secondary energy (electric power) and primary energy, respectively. Therefore, the two accounts play separate roles. However, intrinsically, it is difficult to separate primary energy from secondary energy. In fact, as the function of the two accounts to promote new energies has been strengthened, there have been problems of their functions becoming redundant in recent years. The two accounts were thus merged in FY2007 into the Special Account for Energy Measures.

### Reform of energy taxes and fiscal accounts toward a low carbon economy

What are conditions and issues for reforming Japan's energy tax system toward one that promotes a sustainable low carbon economy? One way to realize such a tax system is to change the tax system so that it encourages low carbon energy use. This means changing the tax base from the amount of energy sources to the amount of carbon contained in them. In fact, as already mentioned, the carbon tax was introduced to reform the existing energy taxes with the name of an "anti-global warming tax".

Energy is a good with a considerably low price elasticity of demand at least in the short run. A substantially high tax rate is thus needed in order to induce energy conservation. However, high rate taxation can significantly increase the burden not only on firms as energy users or sellers, but also on consumers, which makes it politically less feasible. Therefore, with social acceptance taken into account, taxing at a low rate and limiting the use of the tax revenue to the promotion of low carbon energies can be one option to achieve the minimization of impacts on general consumers and an increase in the capacity of the tax system to induce low carbon activities.

Among Japan's seventeen Special Accounts, there has not been any account specialized in low carbon energies, which should be noted in discussing the energy tax system and reform of related accounts toward low carbon activities. The subaccounts for road improvement, river improvement, and harbor improvement of the Special Account for Infrastructure Development, which is connected to public projects, and do not have a mechanism for executing projects that emphasize the protection of the environment or the promotion of low carbon energies. The energy taxes used to be special-purpose revenue sources; that is, the use of the revenue from these taxes was pre-designated. However, today, the revenues from the energy taxes first become general revenues, retaining no obvious mechanism of special-purpose revenues on the surface, as shown in Figure 10.2. At the same time, it is true that, in the end, little change was made to the administrative mechanism of relevant special accounts which are "expenditure windows", as mentioned earlier.

Based on the discussions thus far, let us specifically examine several issues on the direction of a possible comprehensive tax reform toward secured energy sustainability. First, funds for measures for structural improvement of energy supply and demand in the Subaccount for Energy Supply and Demand of the Special Account for Energy Measures get separated and become a source of funds for the "Special Account for Promoting a Low Carbon Economy (tentative name)". That is, the existing Subaccount for Energy Supply and Demand provides funds for ensuring the security of petroleum resources. To maintain consistency, it is appropriate to reallocate funds for structural improvement of energy supply and demand, which are currently used mainly for energy conservation projects or the development of petroleum-substituting energies, to the promotion of low carbon energies.

Second, funds for diversified electric power generation under the Subaccount for Promoting the Development of Power Plants of the Special Account for Energy Measures get separated and become a source of funds for the aforementioned Special Account for Promoting a Low Carbon Economy. At the same time, the Subaccount for Promoting the Development of Power Plants provides funds for the implementation of measures for power plant locations to promote stable supply of electric power.

Third, issues surrounding the promotion of low carbon activities through the taxation framework related to transportation fuel must be considered. Seeing a full-scale conceptual plan to reform transportation fuel taxes into carbon taxes as an issue for a long-term examination, this chapter attempts to consider a case where the promotion of low carbon energies is conducted through the use of revenues from the Gasoline Tax. Though the basic tax rate of the Gasoline Tax is 28.7 yen/l,

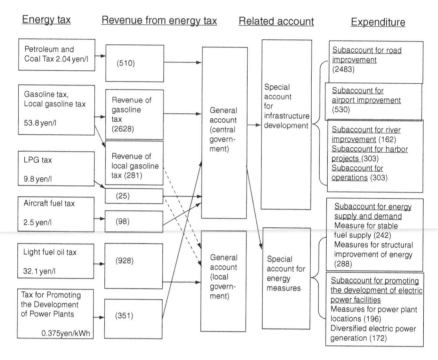

*Figure 10.2* Energy taxes and expenditures from related accounts (after 2007) (billion yen, budget base: FY2009) (source: created by the authors based on data from Ministry of Finance).

an additional 25.1 yen/l has been applied since 1979 to further increase the source of funds for road projects, making the total provisional tax rate 53.8 yen/l. This chapter hypothesizes the use of the 25.1 yen/l added to the basic tax rate as a source of funds for the Special Account for Promoting a Low carbon Economy.[9]

Figure 10.3 shows a reform proposal wherein the Special Account for Promoting a Low Carbon Economy is created by integrating revenues from the Petroleum and Coal Tax, the Tax for Promoting the Development of Thermal power plants, and the Gasoline Tax and budgets related to the promotion of a low carbon society which are scattered within the central government. Based on the FY2009 budgetary standards, the sum of 283 billion yen from the funds for measures for structural improvement of energy supply and demand under the Subaccount for Energy Supply and Demand, 172 billion yen from the funds for diversified electric power generation under the Subaccount for Promoting the Development of Power Plants, about 250 billion yen from the Anti-Global Warming Tax which was introduced in October 2012, and 1226 billion yen from the National Gasoline Tax (the amount from the provisional tax rate minus the amount from the basic tax rate) becomes 2609 billion yen. This is assumed to be the revenue source of the Special Account for Promoting a Low Carbon Economy.

The Petroleum and Coal Tax and the Tax for Promoting the Development of Power Plants are managed by the Ministry of Economy, Trade, and Industry; the Gasoline Tax is managed by Ministry of Land, Infrastructure, Transport, and Tourism. By placing under a unified management the budgets related to the promotion of a low carbon society, which are separately managed by different ministries, and making them a revenue source of the Special Account for Promoting a Low Carbon Economy, overlapping funds or inefficient management can be prevented and can thus contribute to the integration of environmental policies.

This type of plan for restructuring revenue sources related to the promotion of a low carbon economy will become a driving force in the future in making the entire national fiscal measures inclined toward low carbon activities.

## Summary and conclusion

This chapter discussed the current state of the energy tax system and related fiscal accounts and their issues from the standpoint of a shift toward a low carbon economy and examines how the energy tax system should be reformed and how the relevant fiscal accounts should be restructured in order to realize such an economy.

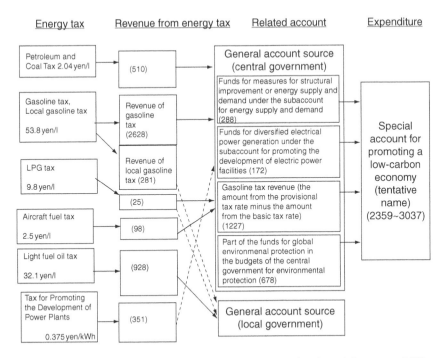

*Figure 10.3* Conceptual plan for restructuring the energy-related special accounts (billion yen, budget base: FY2009) (source: same as Figure 10.2).

To this end, we considered funds managed separately under different accounts with different characteristics that can be used for promoting a low carbon economy and proposed the creation of an account that integrates and manages them. As a realistic solution, this chapter proposes the establishment of the "Special Account for Promoting a Low Carbon Economy (tentative name)" and a conceptual plan to make the acting energy tax system and relevant accounts inclined toward the promotion of low carbon activities. The revenues of this special account will be flexibly used to promote the low carbon economy such as fields of renewable energy, energy saving, forestation, and carbon offsetting, thus eventually achieving tax revenue neutrality by lowering income and corporation taxes with increasing carbon tax rate. This is consistent with double dividend effect; that is increasing environmental welfare due to lower pollution externalities and economic welfare from a smaller deadweight loss of the tax system (Gouldon, 1995).

It is pointed out that the current fiscal framework, in which revenues from energy taxes is treated as special-purpose revenue sources for such issues as the construction of road and steam power generation has not been changed substantially. It is also pointed out that this conceptual plan shows the significance of reorganizing the energy tax system, which has long been vaguely positioned as special-purpose revenue sources or special-purpose taxes, and that the plan is sufficiently feasible within the current budgetary system.

Today's measures toward a low carbon society are diverse and complex, and any solution requires strong coordination among different policy fields and actions overarching them. Policies on energy and the environment become effective through integration with policies on energy, transportation, industry, science and technology, trade, regional issues, and national land planning. In this context, a fiscal account aimed at realizing a low carbon economy is expected to advance policy integration and environmental governance that promotes a sustainable low carbon economy, rather than the interests of individual government agencies.

## Notes

1 Under the FY2003 tax system revision, the Petroleum Tax became the Petroleum and Coal Tax; the target of the taxation was extended to coal.
2 According to Hotelling (1931), a decreased (increased) rate of a rise in the net price, i.e., the net royalty leads to an increased (decreased) speed of the extraction and depletion of an exhaustible resource.
3 Hotelling (1931) discusses monopolistic extraction and a severance tax on firms extracting petroleum, gas, and minerals as a way to control the depletion of resources.
4 Among the energy taxes the Light Fuel Oil Transaction Tax is a local tax. Also, 10 percent of the Gasoline Tax, 50 percent of the Petroleum Gas Tax, and 15 percent of the Aircraft Fuel Tax are transferred to the local governments.
5 Some researchers view the subsidy to the coal industry as a government intervention responding to pecuniary externalities caused by the petroleum industry to the coal industry (Yokoyama 1993).

6 In 1980, the Act on the Special Account for Measures Concerning Coal and Petroleum was revised, and the account became the "Special Account for Measures Concerning Coal, Petroleum, and Petroleum-Substituting Energies". At the same time, funds in the new special account started to be allocated to the development of new energies.

7 The taxes related to transportation fuel became part of the general revenue in April 2009 instead of the traditional revenue designated for road projects.

8 Western European countries have increasingly introduced carbon taxes levied on energy uses as a countermeasure to global warming since the 1990s. There are various types of relationship between a carbon tax and the existing energy taxes in European countries. With the introduction of a carbon tax, the existing energy taxes were reduced in Finland and Sweden, were partially increased in Norway, and were left unadjusted in the Netherlands.

9 In fact, there are opinions that the added tax should be made an environmental tax (see the Government Tax Commission 2002).

# References

Dasgupta, S. and Mitra, T. (1983), Intergenerational Equity and Efficient Allocation of Exhaustible Resources, *International Economic Review* 24(1), pp. 133–154.

Goulden, L. H. (1995), 'Environmental Taxation and the Double Dividend: A Reader's Guide', *International Tax and Public Finance*, 2(2), pp. 159–183

Government Tax Commission (2002), Report on the amendment of Taxation System in 2003 [in Japanese], Cabinet Office.

Hartwick, J. (1977), International Equity and the Investing of Rents from Exhaustible Resources, *American Economic Review* 67(5), pp. 972–974.

Hotelling, H. (1931), The Economics of Exhaustible Resources, *Journal of Political Economy*, 39(2), pp. 137–175.

Ministry of Finance (various years), *Zaimu Tokei (Budget Statistics)* [in Japanese], Tokyo: Ministry of Finance Printing Office.

Petroleum Association of Japan (1980–2012), *The Handbook of the Petroleum Tax System* [in Japanese], Tokyo: Petroleum Association of Japan.

Pigou, A. C. (1920), *The Economics of Welfare*, London: Macmillan.

Shimizu, S. (2002), The Special Account for the Development of Power Plants and Electric Power Deregulation [in Japanese], *Public Finance and Public Policy* 30, pp. 26–33.

Swerling, B. C. (1962), *Current Issues in Commodity Policy*, Princeton, NJ: Princeton University Press.

Takahashi, M. (1972), Energy and Nuclear Power [in Japanese], *Denryoku Keizai Kenkyu*, Central Research Institute of Electric Power Industry, No. 2, pp. 1–75.

Yokoyama, A. (1993), The Energy Tax System and Environmental Tax [in Japanese], H. Ishi, ed., *The Environmental Tax: The Reality and Mechanism*, Tokyo: Toyo Keizai, pp. 47–62.

Zaisei Chosakai (Fiscal Research Group) (1990–2009), *The National Budget* [in Japanese], Tokyo: Hase Shobo.

# 11 Challenges for low carbon policy and fiscal reform in Korea

*Oh Sang Kwon*

## Introduction

Korea is a large $CO_2$-emitting country. Although Korea is not one of the countries with the commitment to limit $CO_2$ emission under the Kyoto protocol, the Korean government has voluntarily suggested 30 percent reduction from the 2020 BAU $CO_2$-emission level. A law, the Basic Act on Green Growth was introduced in February of 2009, and the Low Carbon, Green Growth Policy based on the law that aims at boosting the economy while coping with the growing concern about climate change issues was implemented. Among the policy instruments that the Korean government can choose from to achieve the goal, two quantity oriented policies, the Target Management System (TMS), and the Emission Trading Scheme (ETS) were selected. TMS was introduced in 2011 and is an interim policy toward ETS. ETS is supposed to replace TMS in 2015 and will function as the main policy instrument of inducing $CO_2$-emission reduction.

There is no official action of reforming the current energy tax system yet. Korea currently has a multipurpose energy tax system. The overall tax rates on energy products are rather high compared to those in other countries, but the tax rates are not directly related to the pollution or $CO_2$ contents of energy products. Hence, it is being discussed whether it is necessary to reform the current energy tax system and introduce a more environment-friendly one. It is also a concern how to harmonize the new tax system with the already existing quantity oriented policy instruments, if it is introduced.

In the following, we briefly review $CO_2$ emissions and the price structure of $CO_2$ in Korea. Next, the Low Carbon, Green Growth Policy of the Korean government is introduced and its main instruments to reduce $CO_2$ emission are explained. Several studies that estimated the possible impacts of reforming the current tax system are reviewed. Finally, we simulate the impacts of revising the current energy tax system and making the tax rates more consistent with $CO_2$ contents of energy products under several different scenarios.

## Carbon emission in Korea

Korea, as of 2011, is the ninth largest $CO_2$-emitting country in the world. Moreover Korea's growth rate of $CO_2$ emission is higher than the world average. The growth rate over the period 1990–2009 was 124 percent while that of OECD total over the same period was only 8 percent (Table 11.1). Over the same period Japan recorded only a 2.7 percent increase rate.

Compared to other countries the electricity sector occupies a very large share of $CO_2$ emission in Korea. Almost 49 percent of total Korean emission stems from the electricity and heat production sector while the share in OECD total is only 39 percent, which implies that the Korean economy may depend heavily on electricity (IEA, 2011).

The Korean economy, in general, is $CO_2$ intensive. The figures in the parentheses of Table 11.1 show that $CO_2$ per GDP measured in US dollar is 0.68 kg for Korea in 2009, which is far higher than that of Japan, 0.22, and that of OECD average, 0.41. The Chinese economy is more $CO_2$ intensive than the Korean economy, but the difference between those two economies is getting smaller.

Energy prices, and thus the price of $CO_2$, may be the most important fact affecting $CO_2$ emission in each country. Relatively high share of electricity in total $CO_2$ emission suggests that electricity price might be relatively low in Korea. Korean electricity price measured in US dollar is less than half of that of Japan or OECD Europe (Table 11.2). Although most primary energies are imported from foreign countries, Korean gasoline price relative to that of other countries has a figure very different from that of electricity price. Table 11.2 shows that the price of unleaded regular gasoline in Korea is almost as high as that in Japan, and is far higher than that of OECD average. The table suggests that each energy product is under quite different tax/subsidy rate in Korea. Electricity may be substantially subsidized and, hence, has a lower price in Korea.

Figure 11.1 shows revenues from environmentally related taxes measured in percentage of GDP. Korea belongs to the group of countries where the environmental tax rate is relatively high. It is shown that almost two-thirds of the environmental tax is levied on the energy sector while most of the other one-third is levied on motor vehicles.

What are the components of Korean energy prices? Prices of oil or petroleum products are not directly controlled by the government anymore. Refineries have been setting their prices freely since 1997. However, the government may still control those prices at least indirectly by imposing taxes. The consumer price of oil products and LPG is the sum of crude oil import price, refining costs and margins, marketing costs and margins, and taxes and charges. The last components are controlled by the government.

The structure of electricity and gas prices are quite different from that of oil or petroleum products. Those two products are mostly supplied by public companies that constitute natural monopoly industries. Privatization has occurred in the electricity industry but only in the generating sector. Wholesaling of electricity is still managed by a public company, the Korea Electric Power Corporation (KEPCO). In the markets for electricity and gas the government still plays an explicit role in price determination.

Table 11.1  $CO_2$ emission (million tons) and $CO_2$ per GDP (kg per US dollar)

| | 1990 | 1995 | 2000 | 2005 | 2009 | % change (1990–2009) |
|---|---|---|---|---|---|---|
| World | 20,966.3 (0.86) | 21,791.6 (0.8) | 23,492.9 (0.73) | 27,188.3 (0.74) | 28,999.4 (0.73) | 38.3 (−15.1) |
| Annex I Parties | 13,908.1 (0.7) | 13,179.2 (0.61) | 13,761.9 (0.55) | 14,149.8 (0.51) | 10,236 (0.46) | −26.4 (−34.3) |
| Non-Annex I Parties | 6,444.4 (4.72) | 7,913.9 (4.47) | 8,905.9 (3.57) | 12,074.5 (2.84) | 14,972 (2.42) | 132.3 (−48.7) |
| Non-OECD total | 9,194.8 (2.26) | 9,414.1 (1.94) | 10,033.8 (1.69) | 13,168.7 (1.68) | 15,939 (1.59) | 73.3 (−29.6) |
| OECD total | 11,157.6 (0.55) | 11,679 (0.52) | 12,633.9 (0.48) | 13,055.6 (0.45) | 12,044.7 (0.41) | 8.0 (−25.5) |
| Korea | 229.3 (0.81) | 358.6 (0.87) | 437.7 (0.82) | 467.9 (0.7) | 515.5 (0.68) | 124.8 (−16.0) |
| Japan | 1,064.4 (0.26) | 1,147.9 (0.26) | 1,184 (0.25) | 1,220.7 (0.25) | 1,092.9 (0.22) | 2.7 (−15.4) |
| China | 2,211.3 (4.97) | 2,986.1 (3.77) | 3,037.3 (2.53) | 5,062.4 (2.65) | 6,831.6 (2.33) | 208.9 (−53.1) |

Sources: IEA (2011).

Notes

The figures in parentheses are $CO_2$ emissions per GDP.

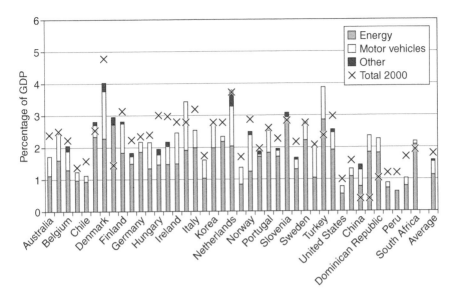

*Figure 11.1* Revenues from environmentally related taxes in percentage of GDP (sources: OECD/EEA Database, 2011, www2.oecd.org/ecoinst/queries/index.htm).

Prices of coal products are also affected by the government policy. The government sets price ceilings of some coal products but subsidizes the gap between production costs and sale prices (Park, 2011).

The differences in industry structure and government tax/subsidy systems imposed on energy products results in different tax/subsidy rates for each energy product (Table 11.3). In general, energy prices in Korea contain 11 types of taxes and subsidies. A consumption tax is imposed on heating oil and industrial oil products. There is a transportation tax imposed only on transportation oil. The current energy tax includes an education tax as well, equivalent to 30 percent of the sum of consumption tax and transportation energy tax. There is a local transportation tax which is 15 percent of the sum of consumption tax and transportation tax. An import levy is imposed on all petroleum products and LNG. Finally, there are several fees and charges.

*Table 11.2* Electricity and regular unleaded gasoline prices

|  | *Electricity (US$/toe, 2008)* | *Gasoline (US$/liter, 2010)* |
|---|---|---|
| OECD | 1,546 | 0.891 |
| OECD Europe | 1,710.8 | – |
| Korea | 699.7 | 1.479 |
| Japan | 1,619.6 | 1.515 |

Sources: IEA (2009b, 2010)

Table 11.3 The structure of energy prices and taxes (won per unit)

| | Gasoline | Kerosene | Diesel | B-C | LPG | | LNG | Briquette | Electricity | | Heat |
| | | | | | Propane | Butane | | | Household | During night time | |
|---|---|---|---|---|---|---|---|---|---|---|---|
| Import duty | crude oil 1%, petroleum products 3% | | | | | | 2% | | | | |
| Consumption tax | 529 | 90 | | 17 | 20 | 275 | 60 | — | — | — | — |
| Transportation energy tax | — | — | 367.5 | — | — | — | — | — | — | — | — |
| Education, traveling (local) tax | 216.9 | 13.5 | 150.7 | 2.6 | — | 41.3 | — | — | — | — | — |
| VAT (10%) | 158.4 | 99.4 | 142.5 | 74.1 | 115 | 150.4 | 94.6 | — | 11.8 | 5.5 | 73 |
| Import levy | 16 | 16 | 16 | 16 | — | — | 24.2 | — | — | — | — |
| Fees, charges | 0.43 | 0.43 | 0.43 | 0.43 | 4.527 | 4.9 | — | — | — | — | — |
| Sales charges | — | — | — | — | — | 62.28 | — | — | — | — | — |
| Sum (w/o import duty) | 920.8 | 219.4 | 677.1 | 110.1 | 139.5 | 533.4 | 183.7 | — | 16.1 | 7.5 | 73 |
| Share in price | 49.8% | 17.9% | 41.0% | 13.3% | 6.9% | 35.7% | 17.7% | 0% | 12.0% | 12.0% | 9.1% |
| Emission intensity (TC/TOE) | 0.78 | 0.81 | 0.84 | 0.87 | 0.71 | 0.71 | 0.64 | 1.01 | 0.0 | 0.0 | 0.0 |

Source: Park (2011).

The consumption tax is imposed mainly out of environmental concerns while the transportation tax is imposed due to concerns about traffic congestion. The education tax is a source of the government's education budget. The local transportation tax revenue is used as a source of local infrastructure budget. The main purpose of imposing import levies is enhancing energy security. Fees and charges are imposed for such issues as quality inspection, and safety control. Thus Korea has currently a multipurpose energy tax system.

The share of tax in total energy price is the highest for the transportation sector, which reaches almost 50 percent. For the energy used for heating and industrial purposes, the proportion of the tax in total price depends on energy type; 17.9 percent for kerosene, 13.3 percent for heavy oil, 6.9 percent for LPG, 17.7 percent for LNG, 0 percent for briquette, 12.0 percent for electricity, and 9.1 percent for heat.

A criticism of the current tax system is that there is no direct relationship between energy's tax rate and its emission (e.g., $CO_2$) intensity because of the multipurpose nature of the current tax system. For instance, in Table 11.3, coal whose $CO_2$ intensity is very high, has a zero energy tax rate. The tax rate on diesel, which has a higher $CO_2$ intensity than gasoline, is lower than that of gasoline.

Improving international competitiveness of the economy is another concern of the current energy tax system. Relatively low tax rates are imposed on the energy products used heavily in industrial or commercial sectors.

There is a cost-sharing issue as well. The price of petroleum, which is used as a fuel for electric generation, is higher than those in other countries, but that of electricity is relatively low, which means that there could be some cross-subsidies provided to electricity.

## Low Carbon, Green Growth Policy

Low Carbon, Green Growth Policy is the main policy of the Korean government aiming at boosting the Korean economy while coping with the growing concern about global environmental issues. The Policy is based on the Basic Act on Green Growth, which was introduced on February 25, 2009 and identified four elements of green economic growth; green technology as a new growth engine, green energy paradigm, contribution to the global community, and improvement in the quality of life. The Policy contains three national strategies and ten key policy agenda, and will be implemented for the following five years supported by several financial measures to promote green investment. The Presidential Committee on Green Growth also has been launched to develop strategies and implement the policy agenda.

In August, 2009 the Korean government proposed, as a part of the Green Growth Policy, three separate $CO_2$ emission limits as potential options for Korea. This was the first step that the government provided the stage for public debate on the country's future $CO_2$ consumption. The government's baseline scenario incorporates the trend of economic growth, energy prices, change in energy efficiencies, population growth, and change in industrial structure. The scenario predicted that total $CO_2$ emissions will increase from 594 million tons in 2005 to 888 million tons in 2030.

The national cuts that the Korean government presented under the BAU scenario were 21, 27 and 30 percent reductions from the 2020 emissions, and those cuts amount to an 8 percent increase, a freeze and a 4 percent decrease from the base year, 2005, respectively. The government eventually chose the 30 percent reduction as the target. It would be a voluntary reduction of $CO_2$ emission since Korea currently does not have an obligation of mandatory reduction under the Kyoto or Post-Kyoto protocol agreements. The reduction goal was imposed on each individual sector, and reductions in transportation and commercial housing sectors are strongly emphasized.

There could be several policy instruments that the Korean government can choose to achieve the goals. The Target Management System (TMS) is the policy of choice. TMS introduced in 2011 is based on the Basic Act. A total of 497 firms, each of which records annual emissions of $CO_2$ more than $125,000\,tCO_2$ (4,236 sources each of which emitting more than $25,000\,tCO_2/$ year), are controlled by the system. Those firms together occupy more than 60 percent of national emissions. It is scheduled that every firm emitting more than $50,000\,CO_2/$year will be covered by the system in 2014, and the proportion controlled by the system will reach more than 90 percent in that case. Currently thirty-six power plants, fifty-one transportation and building companies, 355 manufacturing firms, twenty-eight agricultural and livestock companies, and twenty-seven waste management companies are controlled by the system. Each firm is obliged to submit an emission record, and the target reduction is assigned to the firm based on its four-year historical emission reported. Each firm has to submit an action plan for reduction.

Under the TMS neither trade nor banking/borrowing of reduction is allowed. The system has a form of command and control type policy with very light penalty of violating the reduction target. The system allows a limited flexibility to the firms by imposing reduction targets on firms not on sources. The government has been investing in constructing infrastructure for measuring, reporting, and verifying the reduction.

TMS is an interim policy toward the emission trading scheme (ETS). Again, the Basic Act provides the legal foundation of ETS. The law for the emission trading scheme itself was finally approved by the Parliament in May of 2012. The system is scheduled to be launched on January 1, 2015. Every source emitting more than $25,000\,tCO_2/$year (every firm with more than $125,000\,tCO_2/$year) will be included into the ETS. It is anticipated that most of the initial permits will be grandfathered.

According to the law, ETS will be revised every five years. The Permit Distribution Committee chaired by the Minister of Strategy and Finance will be installed. Permits will be distributed based on the firm's emission record, international trade intensity, emission intensity, and consideration of equity among firms or sectors. For the first five-year stage of the scheme most of the permits will be distributed free. For the second stage, the proportion of free distribution will be determined by the President incorporating the global trend and accomplishments of the first stage. Banking and borrowing may be allowed, but a penalty substantially heavier than that of the current TMS will be imposed.

The law states that firms as well as the government may ask adjusting for emission allowances if unexpected significant changes in allowance costs occur. A comprehensive system of financial assistance, tax assistance, and subsidy to the firms will be also made. There are other provisions on such issues as early reduction, offsetting, and linkage with foreign permit markets.

## An analysis of green tax reform

The major policy instruments that the Korean government has chosen to induce a reduction in national $CO_2$ emission are the quantity-based policies such as TMS and ETS. Nevertheless, the Basic Act still states that the national tax system must be reformed to be more environment friendly. There has been no official action yet to reform the energy tax system and to make the tax rates more consistent with energy products' $CO_2$ intensities. However, a substantial number of government and individual researchers have conducted empirical studies on the possible impacts of introducing a $CO_2$ tax system (e.g., Shin *et al.*, 2010; Kim and Kim, 2010). Most of the works have analyzed the "costs" of the policy reform by estimating the GDP loss or changes in economic growth rates that will be caused by the policy reforms considered. They found that the costs are not so high if the tax or permit revenue is recycled. They employed either an I–O (input–output) or a CGE (computable general equilibrium) model for the analysis.

It is being discussed whether a $CO_2$ tax system has to be introduced or not in Korea mostly by researchers. However, there are very few studies that discussed the direction of reforming the system. Mainly total cost of the potential policy reform was analyzed. We, thus, analyze the impacts of introducing a $CO_2$ tax system under four different scenarios explained below and compare their potential impacts.

Scenario 1 is the case where 10-euro/$tCO_2$ tax is imposed on crude oil, coal, and natural gas which are primary energy sources, in addition to the current energy tax rates shown in Table 11.3. We analyze the impacts of implementing Scenario 1 on the Korean economy's 66 sectors using an extended input–output table of 2007. The reduction in $CO_2$ emission calculated under Scenario 1 is the target level of reduction applied to all the other scenarios. Scenario 2 assumes that the current energy tax rates on the intermediate uses are increased so as to achieve the same reduction as Scenario 1. Under Scenario 3, a new downstream $CO_2$ tax is introduced. That is, additional taxes are imposed on the secondary energy products based on their $CO_2$ contents. Again the additional tax rates are chosen so as Scenario 3 achieves the same $CO_2$ reduction as Scenario 1. Finally, Scenario 4 imposes only a downstream $CO_2$ tax, and the new tax replaces the current multipurpose energy tax.

An input–output modeling introduced by Metcalf (1999) is used for the analysis. Under Scenario 1, 10-euro/tCO$_2$ tax is imposed on crude oil, coal, and natural gas. All coefficients in the I–O table are transformed into carbon units using the IPCC carbon emission coefficients and the Energy I-O Table of Korea.

The national account of the I–O table implies that

$$(I-A')P_0=V \tag{11.1}$$

where $I$ is an N-identity matrix, $A$ is an N-matrix with elements $a_{ij}$, $P_0$ is an N-vector of industry prices, and $V$ is the N-vector of value-added ratios. The system can be solved for the price vector:

$$P_0=(I-A')^{-1}V \tag{11.2}$$

Now let $t_{ij}$ be a unit tax on the use of primary energy $i$ by industry $j$ imposed due to the new carbon tax system. Then the new system under the carbon tax can be solved as:

$$P_1=(I-B')^{-1}V \tag{11.3}$$

where $B$ is an N-matrix with elements $(1+t_{ij})a_{ij}$. $P_1$ is the new price vector, which is the result obtained under Scenario 1. In addition to the usual I–O analysis assumption of fixed input coefficients, the model assumes that (1) all factor price increases are passed forward to consumers, and (2) domestic and foreign goods are sufficiently different so that the price of domestic goods can adjust following changes in factor prices (Metcalf, 1999).

An I–O analysis per se does not provide any direct information on the reduction of CO$_2$ emission since its price model assumes that all quantity indices are fixed. That is, an I–O analysis derives the change in commodity prices induced by carbon taxes but assumes that all quantities do not change despite the price changes. Thus we use an estimated demand system of Kwon and Heo (2011), and combine their elasticity estimates with the I–O price model to derive the quantity changes and thereby the carbon emission changes of the economy. Our analysis is conducted along the following steps:

Step 1: assume a rate of carbon tax and derive the change in commodity prices using the I-O table.
Step 2: apply the estimated price changes into the demand system and derive the changes in each consumption quantity and carbon emission.
Step 3: repeat steps 1 and 2 until the change in total carbon emission meets the target level.

The impacts of Scenarios 2, 3, and 4 are also analyzed following the three steps. It is estimated that Scenario 1 will reduce the total CO$_2$ emissions by 4.4 percent.

The tax rates for Scenarios 2, 3, and 4 are adjusted so that they achieve the same reduction as Scenario 1.

The results on selected industries are summarized in Table 11.4. The table shows that the impacts on each individual industry become quite different when different scenarios are introduced. The prices of fossil-fueled and other power generations will increase substantially under every scenario though.

The last row of Table 11.4 shows the additional tax revenue required to meet the 4.4 percent target reduction under each scenario. An upstream $CO_2$ tax, a tax on primary energy sources, imposed under Scenario 1 induces $CO_2$ reduction with the minimum tax revenue. A downstream $CO_2$ tax that replaces the existing (distortionary) energy tax under Scenario 4 achieves the $CO_2$ emission reduction with almost the same tax revenue as the upstream tax. Increasing the existing (distortionary) energy tax rates under Scenario 2 is the most inefficient policy choice in that it requires the largest tax burden.

Revising the current energy tax system will affect relative prices of commodities and raise carbon price as shown in Table 11.4 depending on the scenario of policy reform. Hence, the policy reform may generate not only the issue of policy cost but also distributional aspects of the cost. Discussions on pollution control through an incentive system have traditionally focused on its efficiency effects such as social optimality or cost-effectiveness. Recently, however, the distribution of policy costs has received increasing attention. First of all, the distributional aspect of a policy itself may be an important issue. Moreover, if the burden of a policy is uneven or unfair, then the policy reform may not be feasible.

The data of Consumer Expenditure Survey are used to analyze the impacts of policy changes on the distribution of consumer expenditure. Household

*Table 11.4* An I-O analysis: price effects on selected sectors (%)

| | Scenario 1 | Scenario 2 | Scenario 3 | Scenario 4 |
|---|---|---|---|---|
| Agriculture, forestry, fishery | 0.89 | 3.40 | 1.71 | 0.58 |
| Coal products | 46.47 | 4.34 | 2.75 | 1.66 |
| Gasoline | 6.40 | 2.25 | 1.78 | 1.54 |
| Diesel | 6.60 | 1.66 | 1.18 | 0.88 |
| LPG | 7.26 | 1.67 | 1.16 | 0.85 |
| Other petroleum products | 7.10 | 3.70 | 8.51 | 12.76 |
| Basic chemical products | 6.91 | 2.76 | 15.86 | 26.11 |
| Iron and steel | 7.97 | 5.91 | 8.76 | 11.07 |
| Hydraulic power | 0.58 | 1.96 | 0.80 | −0.01 |
| Fossil-fueled power | 12.70 | 11.08 | 10.70 | 10.99 |
| Nuclear power | 0.76 | 1.24 | 1.50 | 1.76 |
| Other power generation | 21.21 | 12.01 | 26.09 | 37.34 |
| Urban gas | 7.55 | 0.91 | 0.63 | 0.46 |
| Heat | 7.35 | 14.69 | 8.61 | 4.74 |
| Road transportation | 1.89 | 13.62 | 3.69 | −3.24 |
| Additional tax revenue (billion won) | 10,990 | 27,851 | 17,705 | 11,268 |

Source: own calculation.

incomes are classified into 13 groups as shown in Figure 11.2. It is assumed that the current budget share of each commodity is maintained even if a policy reform is introduced. Figure 11.2 shows the ratios of income classes' additional expenditures to their average total expenditures under the four different scenarios. Class 13 is the richest income class. It is shown that except for the case of Scenario 2 which imposes the largest burden on consumers, the revised tax systems themselves are slightly regressive. However, recycling the additional tax revenue in an appropriate way may transform the systems into progressive ones.

## Summary and conclusion

The main policy instruments chosen by the Korean government to reduce $CO_2$ emission are quantity oriented schemes. The Emission Trading Scheme will replace the Target Management System which is the current quantity regulation scheme in 2015.

Although the Korean government has suggested a road map of $CO_2$ emission reduction, there are still uncertainties associated with the direction of policy reform. The law for the ETS was approved by the Parliament but the actual feature of the scheme is not certain yet.

The government and researchers in Korea are paying attention to tax reform as well, but there is no explicit action yet. There is a general attitude of tax resistance. The current energy tax occupying 6 or 7 percent of national tax revenue may be too important and has too diverse purposes to be revised. Reconciling the $CO_2$ tax system with the existing TMS/ETS is also a concern, if it is introduced.

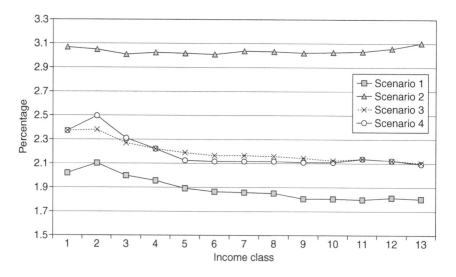

*Figure 11.2* The additional expenditures of various income classes.

Therefore, even if a $CO_2$ tax system is introduced, it is likely that the tax rate is not high. Observers suggest that it may be reasonable introducing a $CO_2$ tax system with very low tax rates and increasing them gradually.

The simulation we implemented suggests that the new tax systems that are more consistent with the $CO_2$ contents of energy products may induce $CO_2$ reduction with smaller distortions compared with the current multipurpose energy tax system.

## References

IEA (2009), *Energy prices and Taxes*, Paris: International Energy Agency.

IEA (2010), *Energy Prices and Taxes*, Paris: International Energy Agency.

IEA (2011), *$CO_2$ Emissions from Fuel Combustion, Highlights*, Paris: International Energy Agency.

Kim, S. R. and J. Kim (2010), *The Design and Economic Effects of Carbon Tax in Korea*, Seoul: Korea Institute of Public Finance.

Kwon, O. S. and D. Y. Heo (2011), "The Incidence of a $CO_2$ Tax in Korea," *Korean Journal of Public Finance* 4(1): 153–180.

Metcalf, G. E. (1999), "A Distributional Analysis of Green Tax Reforms," *National Tax Journal* 52(4): 655–682.

OECD/EEA Database (2011), www2.oecd.org/ecoinst/queries/index.htm.

Park, G. S. (2011), "Issues in Energy Prices," A paper presented in the seminar in celebration of the tenth anniversary of the *Korea Energy Economic Review*, Korea Energy Economics Institute.

Shin, S. C., Y. G. Kim, J. H. Han, and H. Y. Han (2010), *An Assessment of the Feasibility of the Carbon Tax in Korea*, Seoul: Korea Environment Institute.

# 12 Perspectives and challenges of green tax reform in Taiwan

*Daigee Shaw and Pi Chen*

Mitigation of man-made climate change hinges on mankind's reduction of fossil energy consumption and substituting low carbon goods and technologies for high carbon ones. These actions will require and lead to an overhaul in the modes and style of production and consumption. These actions, however, will not happen automatically, i.e., without a price charged on carbon emissions. An emissions tax or other instruments with the same pricing effect may be the most effective way for pricing emissions.

Pricing carbon emissions, however, poses great challenges in both welfare enhancement and welfare redistribution to the contemporary generation. Hence the pace of mitigation and of a green tax reform has been staggering in many countries and, in particular, Taiwan.

This chapter investigates the perspective and challenges facing mitigation and the green tax reform in Taiwan. Some welfare and redistribution implications of a carbon emissions tax universal to many other countries can also be drawn. We outline Taiwan's status of carbon emissions and energy consumption in the first section. It also briefs on Taiwan's mitigation and energy policies and their effects on Taiwan's past and future emissions. The second section introduces some initiatives of a green tax reform in Taiwan. These initiatives have faced great barriers and will continue to do so. The third section explores the potential barriers and challenges facing mitigation and a green tax reform in Taiwan. The final section concludes this chapter with discussions on supplementary policies necessary for mitigation and a successful green tax reform.

## Carbon emissions and the green policies in Taiwan

### Energy consumption and carbon emissions

Taiwan's carbon emissions both in total and in per capita terms have more than doubled since 1990. Its carbon emissions from fuel combustion were 263 $mtCO_2$ in 2007, nearly 1 percent of world totals, ranking it twenty-second in the world. Emissions per capita were 12.08 $tCO_2$.[1]

The fast growth of emissions was a result of Taiwan's fast growth in fossil energy consumption. Overall energy consumption in Taiwan has been more than doubled since 1990, from $47,592 \times 10^3$ toe to $122,372 \times 10^3$ toe in 2007. Energy consumption per head was 4.81 toe per capita in 2007, also more than doubled since 1990.

Taiwan depends heavily on fossil fuels, with oil making up more than half the total primary energy supply (TPES), with coal coming second at nearly one-third in 2007 as cheaper coal penetrated the electricity market. Natural gas, mostly from imported liquefied natural gas (LNG), and nuclear each provides around 8 percent. Solar, wind power and conventional hydro together accounted for just 1.3 percent of the total primary energy supply in 2007 (Figure 12.1). Emissions per TPES in that year were $2.51$ t$CO_2$/toe.

The jump in emissions and energy consumption is a consequence of fast economic growth. Regulated low fuel prices also contribute to this energy extravagancy. Taiwan's electricity prices (\$0.08/kwh for residential rate in 2008) and post-tax gasoline prices (\$0.83/ltr in 2009) have long been among the lowest on earth.

Energy demand is dominated by the industrial sector, which, in 2007, accounted for 52 percent of the total energy consumed. As a result, the industrial sector made up nearly 50 percent of emissions from fuel combustion while accounting for 28 percent of the total GDP. Buildings and road transportation together made up around 40 percent of fuel combustion emissions.

A few studies have attempted to provide insight into the future path for greenhouse gas emissions in Taiwan. The Environmental Protection Administration (EPA) of Taiwan has developed a business as usual case and several abatement scenarios. Without changes to existing policies and regulations or absent a major

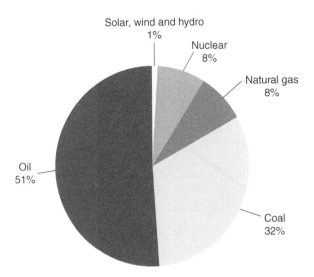

*Figure 12.1* Total energy by fuel (source: *Energy Statistics Handbook 2010*, Bureau of Energy, Ministry of Economic Affairs, Taiwan).

technological advancement, emissions are forecast to rise from $263\,mtCO_2$ in 2007 to 465 mt in 2020 given an annual economic growth rate of 4.5 percent from 2011 through 2020 (the business as usual case in Figure 12.2).

### Energy policy and mitigation measures

The regulated low energy prices in Taiwan plus generally low tax rates on energy products have encouraged the consumption of fossil fuels and pushed up the social discount rate on welfare derived from energy use. A more implicit way to encourage fuel consumption in Taiwan has been to subsidize the usage of auto vehicles in a form that free or low-rate parking on roadsides, public lands and sidewalks is commonly tolerated.

While not a signatory to the United Nations Framework Convention on Climate Change or the Kyoto Protocol, Taiwan has since 2008, responding to the rhetoric of and potential international pressures on mitigation, nominally adopted a climate change policy of "voluntary compliance" to international environmental agreements in its effort to reduce its greenhouse gas emissions. The government announced ambitious goals for carbon emissions reduction: carbon dioxide emissions from fuel combustion in 2020 are set back to 2005 levels (252 mt) and in 2025 back to 2000 levels (215 mt, as compared to 263 mt in 2007).

It is hoped that from 2009 to 2016 an annual 2 percent reduction in energy intensity would be realized and that the energy intensity would, by 2015, decrease by more than 20 percent compared to that in 2005, and by 50 percent by 2025.

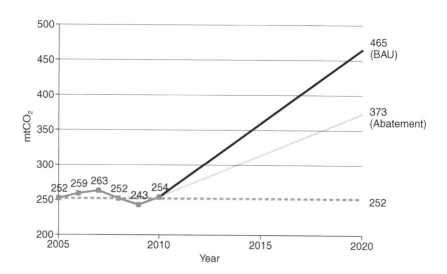

*Figure 12.2* Emissions pathways to 2020 (source: EPA, 2011).

The plan for abating carbon emissions is divided into two areas: developing cleaner energy supply and lowering the growth rate of energy demand. Both the supply and the demand sides place strong emphasis on command and control measures (standards and fuel efficiency) and subsidies. The supply side includes a host of targets, including nearly doubling the share of renewable energy to more than 15 percent by 2025, greater utilization of natural gas, replacement of existing power plants with best available technologies, and a vision of future use of carbon capture and storage (CCS) and clean coal technologies. The weight of nuclear power in energy supply, however, might become much weaker in the wake of the 2011 Fukushima nuclear power plant incident.

The demand side levers include measures for: public sector (government energy reduction plans); residential and commercial sector (low carbon urban planning, improving appliance efficiency standards, efficient lighting programs); transport sector (fuel efficiency standards for private vehicles, mass transport development, intelligent transport systems); industrial sector (energy saving production).

No resolved plan for an emissions pricing mechanism such as a carbon tax or a cap-and-trade scheme is firmly scheduled in the government's mitigation actions.

Taiwan's EPA forecasted a maximum abatement path based on the government's existing mitigation plan. Its result shows a gap between actual emissions and the government emissions target of nearly $120\,mtCO_2$ in 2020 (Figure 12.2) while the measures include only command and control and subsidies but no emissions pricing mechanism.

## Green tax reform initiatives

As Taiwan is having trouble meeting its emissions target, several initiatives for an energy tax act brought up by the Cabinet, the Parliament and some non-government organizations have been debated in public forums. Some reckoned that asking people to pay for emissions through a carbon emissions tax or a cap-and-trade scheme may be more quantity- and cost effective than fuel economy and emissions standards alone.

Chung-Hua Institution for Economic Research (CIER), a think tank in Taiwan focusing on economic policies research, in particular, proposed an energy and emissions tax to be implemented gradually over a ten-year period. The taxes are structured to act as an excise tax on carbon emissions and on consumption of exhaustible energy resources. The tax bases include gasoline, diesel, jet fuel, LPG, fuel oil, kerosene, natural gas, coal and nuclear. The fully implemented total tax rate in terms of carbon emissions is around 25 US$/$tCO_2$e.

Assuming mostly inelastic demand (<1) for the taxed fossil fuels, CIER estimated that in the tenth year after the enactment of the tax, carbon emissions could be diminished by about 46 million tons, roughly 22 percent of the necessary reduction as estimated by EPA for the 2020 emissions goal. It could also collect a grand sum of tax revenue making up 27.8 percent of Taiwan's 2010 grand total tax revenue, surpassing the share of revenues from individual income tax.

The revenue from these taxes was suggested to be recycled by means of subsidies for public transportation, research and development for energy savings
and emissions reductions, and lowering income tax rates. It is also to be spent on
subsidies to low income families for increased energy prices.

CIER's plan, however, according to simulations from some computable
general equilibrium models, may cause some dampening in GDP growth rates in
the first several years, but eventually returns to positive growth in the 2030 time-
line (CIER, 2009).

## Challenges facing the green tax reform: economic growth, distributive justice and fiscal stability

Why has Taiwan not fully committed itself to a more sweeping, more indiscrimi-
nate abating tool like an emission tax after it announced ambitious emissions
reduction goals? A sizable cut in emissions could mean a revolution in commut-
ing fashion, relocation of housing, increases in the production costs and loss of
profits for some businesses, sacrificing some opportunities of consumption and
investment, and so on. So the government and the Taiwanese society may fear
that a serious emissions cut would demand too high a price from Taiwan. The
size and distribution of the cost, in the end, would critically affect the enforcea-
bility of any ambitious emissions reduction.

We find four major tradeoffs generated from emissions cuts and a green tax
reform to be overcome:

* emissions reduction vs. GDP growth (tradeoffs between efficiency and GDP
  growth);
* welfare for future generations vs. welfare for the contemporary generation;
* emissions reduction vs. intra-generational equity (tradeoffs between efficiency and equity);
* a greener tax system vs. fiscal stability.

### Economic growth and intergenerational welfare transfer

Consumers, businessmen and politicians, when facing an emission tax proposal,
sometimes interpret the tax amounts the citizens will have to pay for emitting
carbons as the *social* cost of mitigation. The tax payment itself, however, is *not* a
part of the social cost. The social cost of emissions reduction, moreover, is
usually not as big as the tax amounts paid for emissions. While an emissions tax
costs carbon emitting businesses and consumers, revenues collected by the government from this tax can be recycled to the society for growth-promoting programs or social welfare entitlements. The social cost incurred from an emissions
tax is only the deadweight loss due to a decrease in the net benefit from foregone
production and consumption. And this social cost will need to be evaluated
against the *future* climate benefits.

Two key things blockade the way to a collective choice on emissions cuts, however. First, some of the benefits of emissions reduction are ecological or environmental, but not in terms of the conventional GDP enhancement. But continuing and robust GDP growth is a key factor in job creation. Unemployment inflicts sometimes life-long damage to the unemployed workers. As Taiwan has been confronted by a surge in unemployment rates during the last decade, it may be hard to convince the Taiwanese people that sacrificing GDP growth and the derived additional job opportunities for emissions reduction is worth doing.

Second, while today's citizens would bear a good portion of the expenses of emissions reduction, its benefits will be mostly collected in the future. If we expect that GDP continues to grow in the long run with *and* without a sizable emissions tax today, then an emissions tax system could channel an intergenerational transfer from the poorest generation to the richer generations. This violates such ethical standards as the Rawlsian Maxmin principle.[2]

These two dilemmas may concern today's Taiwanese society and many others. The fact that CIER indicated in its tax proposal that GDP would eventually return to positive growth after a sizable emissions tax may not, in the face of the above two dilemmas, be persuasive enough for the Taiwanese citizens to sacrifice their welfare for the future generations.

A bad-timing issue also comes to bear as the globe is still struggling with the aftershocks of the world financial crisis. Unless strong evidence indicates that an emissions tax would not slow the pace of a healthy recovery, the Taiwanese government may hesitate about any substantial exercise of such a tax.

A serious emissions reduction faces further challenges when potential carbon leaks exist. Global warming is a global public bad. Any emissions reduction in one country can be neutralized by an increase of emissions in another country if there is a lack of international coordination. An emissions tax may force some industries moving to a country whose emissions regulations are even looser than the pre-tax Taiwan's, and thus emitting even more than in pre-tax Taiwan. So the emission tax, it is argued, is no good either to the global climate or to Taiwan's GDP growth in this case.

### *Intra-generational redistribution*

Increase in energy prices due to an emissions cut and an emissions tax would cause income and wealth redistribution not only across different generations, but also within the contemporary generation. An energy-related tax is widely viewed as regressive because, it is believed, energy consumption makes up a bigger portion in the total consumption of low-income households. If this is indeed the case, and if the future generation would get richer anyway, then those who sacrifice the most under an emission tax are today's poorest people.

A further deterioration of the poor's financial position can happen if, as CIER envisioned, with the expected revenues from this tax the government wants, as a means for an economic-efficiency-enhancement tax swap, to lower the rates of the distortionary but progressive income tax.

Taiwan, among many better developed countries, has actually been seeing an enlarging income gap between the rich and the poor in the last decade. A potentially regressive emissions tax thus might also be viewed as bad timing in the equity realm.

### Tax restructuring and fiscal stability

CIER's proposal showed that the revenue from an energy and emissions tax in Taiwan could eventually surpass 2010's share of revenues from individual income tax. This means that not only has the tax base shifted from the direct tax to the indirect tax, but also to a narrower, energy-related base. In addition to its implication on equity as stated, this shift may also bring in an instability factor to the government's revenue. CIER's simulation of tax revenues was based on an inelastic demand for fossil fuels. If, however, in the long run the demand elasticities for fossil fuels get higher, more fossil energy products would be substituted by non-fossil fuels, leading to a decrease in tax revenue. Thus a heavy reliance on the fossil-fuel-based tax as the revenue source could endanger fiscal stability.

## Concluding remarks

An emissions cut action is a correction for the negative externality from fossil fuel consumption. Since it will increase the scarcity of energy and the environment, cutting emissions may also help stimulate technological innovations aiming at overcoming the Malthusian constraint in energy and climate change. With its benefits to be reaped in the future but costs to be paid today, however, it will also cause an intergenerational welfare transfer with a side effect on intragenerational redistribution. To summarize, three objectives commonly valued by most societies – global efficiency, GDP growth and job creation, and equity – may be in conflict with one another in engineering a green tax system and in fighting against climate change (Figure 12.3).

So how much today's citizens will have to pay for this correction, and how the cost will be distributed among the contemporary citizens, is critical for the economic and political feasibility of a substantial mitigation. To find a balance among the necessary mitigation, GDP growth and equity is a difficult task. Clearer guides, however, can be given on how, through a carefully crafted tax recycling scheme, to remedy the potential efficiency and GDP losses, and how to ensure that the population least able to afford an increase in living expenses does not shoulder the brunt of higher energy costs.

The size and the distribution of the abatement cost would not only be affected by the reduction scale and the size of an emissions tax, but also hinge on how the tax revenue is recycled. The revenue from an emissions tax can be rebated to citizens in a lump-sum fashion. Alternatively, as mentioned above, with this revenue the government can engage in a tax swap by simultaneously lowering the rates of some distortionary taxes. This would result in an efficiency

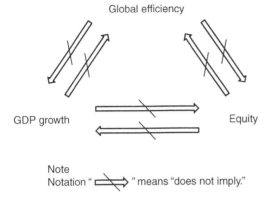

Note
Notation "⟹" means "does not imply."

*Figure 12.3* Tradeoffs between social objectives.

improvement over the lump-sum transfer approach, and a comparatively smaller social cost of emissions reduction.[3]

A lump-sum rebate has its merit with respect to equity, however, if it is channeled to the low-income households. CIER's proposal actually split the rebates of emissions tax revenue into three major areas: lowering income tax rates, subsidizing R&D and transferring this money to the low-income households. To be sure, how to optimize the distribution among them is as difficult as how to find a balance among the three valuable social objectives.

The bottom line is, however: no redistribution should be biased toward greater inequality while this arrangement leaves a foregone chance of improving economic efficiency. A case in point: the grandfathering rule in the cap-and-trade scheme. It is commonly suggested that in a cap-and-trade system the government distributes the emissions allowances to businesses for free according to their historical emissions records. This is simply the same as a lump-sum rebate, with the revenue from an emissions tax, to the businesses having emitted carbon. And this, in general, is worse in the efficiency sense than a productive tax swap and in the equity sense than a transfer to the least able people.

## Notes

1  Bureau of Energy, Ministry of Economic Affairs, Taiwan. Energy consumption and emissions in 2009 were less than those in 2007, perhaps largely due to the great economic contraction since 2008 and the rising energy prices trend. Energy use and carbon emissions might revert to a higher track, even if energy prices remain high, once the economy truly recovers.

2  The Rawlsian Maxmin principle is based on the philosophical work of John Rawls and, to quote Wikipedia's entry on Social Welfare Function, "measures the social welfare of society on the basis of the welfare of the least well-off individual member of society" (http://en.wikipedia.org/wiki/Social_welfare_function, accessed 3 January 2013). Any collective action that enlarges the aggregation of individual members' welfare at the

expense of welfare stagnation or deterioration of the least well-off individual member of society does not, according to the Rawlsian Maxmin principle, enhance the social welfare.

3 This result is sometimes called the "double dividend" of emissions reduction in the weak sense. The first dividend of the double dividend refers to a decrease in environmental risk due to emissions reduction. The second dividend has more than one definition. A stronger sense of it implies that replacing distortionary taxes such as the income tax by the emissions tax will not only reduce the environmental risk but also increase aggregate economic welfare. Whether this tax swap will generate such a dividend is theoretically and empirically controversial in economics literature. The weak form discussed in the context claims that replacing a distortionary tax by the emissions tax, while probably unable to raise aggregate economic welfare, is in the efficiency sense better than a lump-sum rebate. The weaker form of the double dividend hypothesis is logically and empirically better established than the stronger one.

## References

Chung-Hua Institution for Economic Research (CIER) (2009) *A Proposal for the Green Tax Reform* (in Chinese, 綠色稅制之研究), commissioned by the Tax Reform Committee, Executive Yuan, Taiwan.

EPA (2011) *The Action Plan for Emissions Reduction and the Division of Labor for the Related Key Strategies* (in Chinese, 節能減碳行動方案及關鍵策略分工). Available at: http://web3.moeaboe.gov.tw/ECW/reduceco2/content/ContentLink.aspx?menu_id=785 (accessed 3 January 2013).

Ministry of Economic Affairs (2010) *Energy Statistics Handbook 2010*, Bureau of Energy, Ministry of Economic Affairs, Taiwan

# 13 Environmental fiscal reform in Thailand

*Chapika Sangkapitux and Akihisa Mori*

## Introduction

As developing and emerging countries try to move towards low carbon growth, fiscal and legal reforms of their environmental and energy policy frameworks are imperative. This applies particularly for Thailand, whose rapid and energy-intensive economic growth and accelerating urbanization has posed enormous challenges for environmental sustainability and the transformation of energy systems. Thailand has turned into a net energy importing country and is now one of the largest emitters of greenhouse gases (GHG) in Asia. A joint report by the World Bank and the Thai National Economic and Social Development Board (NESDB) stresses the urgent need of the Thai government to step up its efforts towards deploying pricing and fiscal measures and clearly defined, performance-based energy-saving targets to curb the rising energy intensity in the country (World Bank and NESDB, 2011).

Yet, addressing these well-founded calls for fiscal environmental and energy policy reform is not an easy matter, given the various political, institutional, structural and legal barriers that subsequent Thai governments have mostly failed to overcome in their past reform efforts. This chapter aims to review the progress of energy and environmental policies in relation to fiscal reform in Thailand and to draw implications for future policy reform. The literature-based analytical part elucidates the institutional and political context that may enhance or hinder the policy reform efforts in the environmental and in the environment-related energy sector.

The chapter is organized into five sections. Following this introductory section, the second section sketches out the development of the environmental fiscal mechanism. The third section discusses the environmental policy reform with an emphasis on two main issues: first, the political controversy surrounding the draft "Fiscal Measures for Environment Act", and, second, barriers to implementing economic instruments in pollution control, drawing on the case of public waste-water charges. The fourth section sheds light on the environment-related policy reform process in the energy sector. To this end, the context of energy policy and planning is provided, followed by a discussion of successes and failures in energy policy reform, which is exemplified by three particular cases: (1) the oil price subsidy, (2) the adder program for renewable energy promotion and (3) tax incentives for bioenergy promotion, with a focus on bioethanol. The sixth section summarizes the major findings and draws some conclusions.

## Development of the environmental fiscal mechanism

Thailand introduced the environmental fiscal mechanism with the enactment of the Enhancement and Conservation of National Environmental Quality Act 1992 (the 1992 NEQA). The Act gave authority to the National Environment Board (NEB) to designate pollution control areas and environmental protection areas to localities so that local governments take tougher measures. Local governments in turn are obliged to submit a pollution control plan once designated as pollution control areas. The Thai Environmental Fund was established under the Office of Environmental Policy and Planning (ONEP) to provide subsidized loans and grants to companies and local governments for environmental projects in designated areas. While during the initial years the fund was allocated exclusively to such environmental projects, fiscal demand for environmental protection has not been limited to the designated areas. In response, the central government mandated all the provincial governments to submit provincial environmental action plans so that ONEP can allocate the budget to support their environmental projects.

However, government environmental expenditure was not large enough to address worsening environmental degradation. It amounted to only 0.14 per cent of GDP and 0.63 per cent of government expenditure at most (Figure 13.1). As a result of the Asian economic crisis the government placed lower priority on the environment, which led to stagnation in government environmental expenditure and its share in GDP during 1996–2002. Nonetheless, the so-called "People's Constitution" of 1997 that ascribes rights and obligations of local governments to manage and preserve the environment, and the Decentralization Act of 1999 that stipulated municipality's responsibility for solid waste management raised fiscal demand from municipalities.

The Thaksin government (2001–2006) instigated a bureaucratic reorganization in 2002 that included the establishment of the Ministry of Natural Resource and Environment (MoNRE). Besides three departments that had been previously established, MoNRE took over the Forest Department from the Ministry of Agriculture, the Department of Mineral Resources from the Ministry of Industry and newly created departments in charge of groundwater resources, marine and coastal resources, and greenhouse gas management.

The reorganization increased the share of budget allocated to MoNRE, while the budget allocation among the various departments has been significantly changed. The share of budget allocated to the original three departments shrank due to a higher budget allocation to the water, groundwater and natural park departments (Figure 13.1). This reduction became more apparent in 2004 when the Thaksin government started the fiscal transfer to local governments through the Central Fund and the Revolving Fund instead of through ministerial budget, which resulted in a dramatic drop of the budget for ONEP.

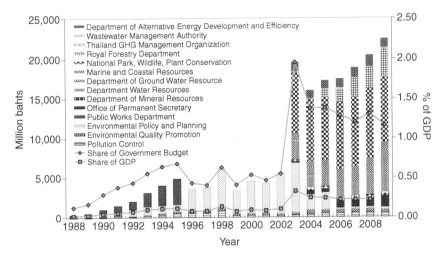

*Figure 13.1* Government environmental expenditure in Thailand (source: author compilation based on *Statistical Yearbook of Thailand*, various years).

In the bureaucratic reorganization in 2002, the Department of Energy Development and Promotion was reorganized to be the Department of Alternative Energy Development and Efficiency under the newly established Ministry of Energy (MoE). The department takes charge of development and diffusion of renewable energy, as well as energy conservation.

## Fiscal reform in Thailand's environmental management: constraints and barriers

Fiscal reforms in environmental management have been implemented in most industrialized countries and were recently also introduced in a number of developing countries, Thailand being one of them. In the following, we will discuss the political controversy surrounding the draft "Fiscal Measures for Environmental" Act (FME Act) in Thailand, followed by a review of barriers to implementing economic instruments for pollution control, drawing on the case of public wastewater charges.

### Political controversy over the draft "Fiscal Measures for Environmental" Act

With support from the Asian Development Bank (ADB) the Office of Fiscal Policy under the Ministry of Finance (MoF) completed the draft "Fiscal Measures for Environmental" Act in 2007. This act is to become an "umbrella" legislation in the form of a Royal Decree allowing environmental protection agencies to apply economic instruments deemed appropriate to control certain forms of pollution. These instruments are proposed as a package of measures in six categories, namely (1) environmental tax, (2) waste management fee, (3) product tax and surcharge, (4) deposit insurance for environmental risk and damage, (5) trading of natural resource use rights or pollution emission rights and 6) subsidies, support measures or other privileges (Pollution Control Department, 2012). As additional measures, an Environmental Tax Fund and an environmental surcharge were introduced. The draft of the FME Act was approved in principle in a cabinet meeting of the then Thai government led by the Democrat Party in October 2010. All affected ministries were requested to take note of the Cabinet resolution and to pass on the result to the office of the Council of State for further consideration (Pollution Control Department, 2012). However, the act has yet to go through the parliamentary process and progress towards this end has been very slow, due to a variety of factors discussed below.

### Communication and information failure

Information sharing and communication among interest groups are crucial prerequisites for the successful introduction of economic instruments (EIs) for improved environmental management. EIs have been suggested by resource economists as more promising alternatives to conventional command-and-control approaches. This is also true for the FME Act which was initiated and strongly supported by Thai resource economists. While a number of interest-group meetings have been organized, the approach was perceived as imposed by its academic supporters on the policy makers and thus failed to convince the various interest groups. Such communication failures have led to adverse impacts, manifested by opposition from both policy makers and business circles, which has undermined the support of the policy initiators.

### Conflict with economic priorities and fear of loss of competitiveness

To pursue economic and social development, official government priorities are geared towards economic growth, poverty reduction and equity in most developing countries, while environmental concerns often come last. As many environmental policies are likely to have a harsh impact on economic objectives by curbing profits of polluting industries and ultimately increasing prices for consumers, governments seeking reelection are often reluctant to support such unpopular policies. The pollution tax, as an example, has not been successfully introduced to many developing countries and emerging economies, due to its perceived negative impact on economic competitiveness.

### Vested interests and power struggles among bureaucrats

This newly proposed act is meant to be the master law requiring close cooperation among various ministries and their line agencies, which makes it even more difficult to go through the national approval legal procedure as compared to other legal initiatives that are based on single agency responsibility. In this particular case, the MoNRE could be seen as a ministry that is likely to have its power reduced, while the MoF as a newcomer in the arena would take benefit from this act. The Minister and a senior executive at MoNRE argued that at least 99 regulations under the draft of the FME Act are conflicting with the current regulatory and legal framework under the various departments of MoNRE (*Matichon* Online, 3 November 2010). In addition, there are several interest groups, such as the corporate sector, that are very influential and have lobbied successfully under the existing regulatory setting.

### *Difficulties in compromising and integrating with the existing regulatory setting*

The existing regulatory setting for environmental management has been criticized for its ineffectiveness and inefficiency in response to the dramatic changes of Thailand's environmental conditions which arguably should be better handled under the newly proposed FME Act. The act is deemed by many decision makers as having some aspects which are not clearly distinguished from the existing ones causing a considerably delay in having the act approved. One example of the unclear boundaries between new and old environmental legislation is the overlap that exists between the already established Thai Environmental Fund and the newly proposed Environmental Tax Fund. Due to the inefficient management of the existing Thai Environmental Fund under MoNRE, the Environmental Tax Fund is proposed under the FME Act to be managed by the MoF providing more flexibility in fund managing schemes. This has raised a concern of overlapping responsibilities and unclear distinction between these two funds by the previous minister of MoNRE (*Prachachart* Online, 4 November 2010). The Royal Decree Committee has suggested the Office of Fiscal Policy under the MoF revise this part of the draft.

### **Barriers to implement economic instruments in pollution control: the case of public wastewater charges**

Prior to 1992, waste water management (WWM) was under the responsibility of the Public Works Department (PWD), a line agency under the Ministry of Interior (MoI), whose duty included the planning and design of wastewater treatment systems. Based on their plans, the central government allocated the construction costs of the facilities before transferring them to local government organizations for operation and maintenance. Three new organizations, namely ONEP, the Pollution Control Department (PCD) and the Department of Environmental Quality Promotion (DEQP), were established in 1992 under the Ministry of Science, Technology and Environment (MoSTE). Subsequently, the responsibility for WWM was transferred from the MoI's PWD to MoSTE as part of a structural reform. Backed by politicians, however, it was not until 1996 that the

PWD handed over the authority and budget to the MoSTE, while PWD initiated construction of wastewater treatment systems in the pollution control areas.

The "Decentralization Act" of 1999 devolved wide-ranging decision-making power and fiscal responsibility to local government organizations, requiring them to take responsibility in the area of public services, including environmental services, such as pollution control. Yet, in reality, most local governments rely on central government support as the initial investments in wastewater treatment facilities tend to be high and cannot be afforded by local resources only (Rammont and Amin, 2010). In addition, the government and ONEP prefer investment in central treatment plants for wastewater and hazardous waste due to expected economies of scale. Owing to the limited budget of government line agencies, most central treatment plants are built in public–private partnerships (Kititasnasorchai and Tasneeyanond, 2000). The central government subsidizes capital investment in the wastewater management sector through two main channels: (1) budgetary allocation through the Budget Bureau and (2) grants and soft loans through the Thai Environmental Fund (Rammont and Amin, 2010).

The polluter-pays principle (PPP) was introduced in Thailand through the 1992 NEQA. The policy of levying charges for pollution control in general and WWM in particular has also been outlined in the Seventh National Economic and Social Development Plan (1992–1996) and it remains in the Eleventh Plan (2012–2016). Yet the actual implementation depends to a large extent on the local government authorities (LGAs) since it is they who are in charge of implementing the national policy towards providing environmental services, including the establishment of wastewater treatment systems, according to the Decentralization Act of 1999. In cases where a facility has been built with a grant from the Thai Environment Fund, the NEB has the power to set the wastewater treatment charge for the area covered by the facility (Kaosa-ard *et al.*, 2008). In all other cases, LGAs are endowed with the authority to implement the wastewater charge (Rammont and Amin, 2010). Yet only few local governments, e.g. in Pattaya and Chonburi, have levied wastewater treatment charges to date, although it would be an important means to recover the operating costs (Kaosa-ard *et al.*, 2008). Barriers to domestic wastewater implementation stem from a variety of factors including insufficient technical knowledge and manpower, vested interests and power struggles among bureaucrats, political sensitivity and lack of public participation. Each of these factors will be discussed below.

*Insufficient technical knowledge and manpower*

The case of Rayong, where the physical structure of a wastewater treatment plant has been built, but still is not operational, provides a prototypical example for technical and human capacity problems. The facility which was constructed by the former Department of Public Works several years ago was technically flawed and unmanageable with the existing manpower, with the consequence that the Rayong city municipality refused to take on its operation (Kaosa-ard *et al.*, 2008). The lack of skilled workers leads to inefficiency and increased maintenance costs. This underscores the persisting problem of overemphasizing "hardware" issues – such as technology transfer – at the expense of "software" issues – such as improvement of human skills and provision of operating knowledge.

## Institutional segregation

Institutional segregation is a widespread phenomenon in the Thai administrative and bureaucratic landscape. Two major authorities share responsibility for water supply services, namely (1) the Metropolitan Waterworks Authority (MWA) which provides services only in Bangkok, Nonthaburi and Samut Prakarn, and (2) the Provincial Waterworks Authority (PWA) which serves all other provinces. While the waterworks authorities are in charge of treating water and distributing it to customers, the wastewater generated from several sources is collected and treated by local government organizations. Due to this institutional segregation of the water supply and wastewater management sectors and lack of cooperation between the Bangkok Metropolitan Administration (BMA) and MWA, the collection of wastewater charges in Bangkok metropolitan area has been repeatedly postponed because the household registration database of the BMA could not be synchronized with that of the MWA (Simachaya, 2009). This mix of unwillingness and incapacity of these two organizations to coordinate water supply and wastewater management services is emblematic for the lack of inter-institutional cooperation and a major impediment for implementing more effective environmental policies and standards.

## Political sensitivity: national and local politicians, NGOs and businesses

As the authority to implement and set charges or fees for wastewater management is held by political and administrative bodies at various levels of government, the charge/fee setting process is often dominated by political objectives and agendas. Most elected officials and staff of LGAs perceive beneficiaries as reluctant to pay for wastewater services and therefore hesitate to charge them (Simachaya, 2009). Provincial local governments encompass two types of officials – those elected by residents (mayors and municipal council members) and those appointed by the government (governors). Obviously, downward accountability is stronger among the elected officials and wastewater charges tend to become an election issue. The case of the BMA is more complex: while feasibility studies are conducted and charge rates and dates of implementation are announced, their application is routinely postponed due to the perceived political risk of enforcing wastewater charges and the major influence of the country's capital on the course of national politics (Rammont and Amin, 2010).

## Lack of public participation and acceptance

The notorious lack of public participation in wastewater management issues has been emphasized by a number of authors and institutions (e.g. World Bank and

NESDB 2008; Rammont and Amin, 2010). This is manifested in both plant construction and charge/fee implementation. The World Bank and NESDB (2008) hold that understanding among communities regarding the importance of wastewater treatment systems is often limited and the public is insufficiently informed and rarely involved in decision-making processes, which triggers complaints and resistance from those communities that are directly or indirectly affected by the construction and operation of wastewater treatment facilities.

## Environment-related fiscal reform in energy sector

Thailand's energy consumption has increased remarkably with the economic recovery process that followed the 1997 Asian economic crisis. The industrial sectors with the highest growth rates include the construction material and the automobile and motorcycle manufacturing, which have a particularly high energy demand. In order to cope with rising energy costs and to reduce the country's reliance on imports, Thai governments have developed policies pursuing three main objectives; (1) securing energy supply through developing new energy sources, (2) improving energy efficiency, (3) conserving energy. Tremendous efforts have also been made to reduce the kingdom's reliance on energy imports.

In order to facilitate all policy measures developed under the guidance of the national long-term energy plans, the well-established Energy Conservation Promotion Fund (ENCON Fund) has been utilized. In April 1992, the Thai government had already enacted the Energy Conservation Promotion Act aiming at enhancing energy conservation discipline and investments in energy conservation measures in factories and other buildings. This went along with the establishment of ENCON Fund under the National Energy Policy Office (NEPO) to support such measures financially. Yet the act also includes a punishment clause targeting the owners of factories, office buildings or condominiums who do not comply with the standards and procedures that are included in various ministerial regulations formulated under the act. Hence, the act is a combination of "carrot and stick", i.e. includes both incentives and sanctions. With start-up capital from the ENCON Fund the MoE has further instigated two funds, namely the Energy Services Company (ESCO) Fund and the Energy Efficiency (EE) Revolving Fund to foster private investments in Renewable Energy and EE projects (Asia-Pacific Economic Cooperation, 2005). This has resulted in growth rates of renewable energy that exceed those of non-renewable energy sources (Table 13.1, p. 236).

### Energy policy and planning

Thailand's energy policies are developed and proposed by the MoE and enshrined in three long-term plans: (1) the Power Development Plan (2010–2030), (2) the 20-Year Energy Efficiency Development Plan 2011–2030 and (3) the 10-Year Alternative Energy Development Plan 2012–2021.

1   The Power Development Plan 2010–2030 aims at ensuring security and appropriateness of the nation's power system, in line with the policies of the MoE to address environmental concerns, energy efficiency aspects and renewable energy promotion.

2   The 20-Year Energy Efficiency Development Plan 2011–2030 aims at a 25 per cent reduction of energy intensity – expressed in the ratio of energy consumption to GDP – in the country within 20 years.

3   The 10-Year Alternative Energy Development Plan 2012–2021 aims at replacing fossil fuels by various alternative energy sources for power plants. Within these ten years, the share of renewable energy is to be increased by 25 per cent.

## Successes and failures in energy policy reform: institutional and political aspects

### Oil price subsidy

Various Thai governments have capped the retail oil price to provide equal access to Thai citizens. Following the first global oil shock in 1973, Thailand established the so-called Oil Fund to hedge domestic oil prices against fluctuations in international price levels. It collects a levy on petroleum from importers and domestic producers, thus accumulating reserves that can be used to subsidize retail prices of gasoline and diesel at times when they pass a certain threshold. As diesel plays a major role in Thailand's transportation sector – accounting for about half of the country's domestic petroleum consumption – particular attention is paid to keeping diesel prices below a certain threshold. An increase would affect Thai citizens through rising food and transportation cost. The diesel subsidy scheme supported by the Oil Fund was in place until 2005, when the diesel price was liberalized (Credit Suisse, 2011).

The Democrat-led Thai government announced a reintroduction of diesel subsidies as part of a pre-election package in December 2010. In April 2011, following several months of high oil prices and the subsequent depletion of the Fund's reserves, the Thai government replaced the Oil Fund's fuel subsidy system with a reduction of the excise tax in order to keep retail prices below 30 THB per litre. Originally announced as a temporary measure, it has since been extended several times. The IMF calculated a revenue loss from the excise tax reduction of 1 per cent of GDP, if it were to be maintained over an entire year (IMF, 2012). As of June 2012, the oil fund ran a deficit of about 21 billion THB, according to the *Thai Financial Post* (14 June 2012). A National Energy Policy Council (NEPC) meeting in mid-2012 resolved to raise the benzene and gasohol sales contribution to the oil fund by 0.50 THB per litre, thus collecting more than 150 million THB. At the same meeting, the NEPC decided to cut the levy on diesel by 0.50 THB to maintain the diesel price at the mid-2012 level (*The Nation*, 24 July 2012).

The most recent one-month extension of the temporary excise tax cut for diesel fuel from 1–30 September 2012, was estimated to cost the government 9 billion THB in lost revenues (*Bangkok Post*, 22 August 2012). A recent IMF study concluded that such tax incentives should be eliminated because they are fiscally unsustainable (International Monetary Fund, 2012). In addition, these measures are also deemed ineffective with regard to providing a more equitable income distribution, since better-off households get the lion's share of the benefits due to their higher rate of energy consumption. Yet, policy makers continue to opt for diesel price subsidy schemes, as their constituencies prefer to enjoy relative price stability of food and transportation. Politicians who oppose such populist measures and aim at introducing fiscally responsible energy policies are quickly replaced, as exemplified by the dismissal of the former Energy Minister, Pichai Naripthaphan, a strong supporter of floating energy prices. By the end of January 2012, a new Minister of Energy was appointed and the plans of floating energy prices have been shelved, while the subsidy scheme has been extended for an unspecified period.

### Adder program for renewable energy promotion

Thailand instigated an adder program to provide incentives for investments in renewable energy in the year 2007. The adder program guarantees power purchasing rates for renewable electricity at specified premium tariffs for a specified time period. Local utilities conclude long-term contracts with eligible providers that are distinguished into Very Small Power Producers (VSPP) (up to 10 MW capacity) and Small Power Producers (SPP) (between 10 and 90 MW capacity). Implementation of the adder program is done by Thailand's three state-owned electricity providers, namely Electricity Generating Authority of Thailand (EGAT), the Metropolitan Electricity Authority (MEA) and the Provincial Electricity Authority (PEA), which purchase electricity from renewable energy generation facilities. The premium price that these small and very small providers get under the program depends on the electricity source and type of technology. To date, about 82 per cent of Thailand's online capacity for electricity from renewable sources is derived from biomass, such as rice husk (Greacen and Greacen, 2012). With regard to signed Power Purchase Agreements (PPAs) as an indicator of future capacity, biomass is slightly surpassed by solar power which accounts for nearly half the capacity under PPAs (Table 13.2).

In developing the adder program, extensive consultations were conducted with the private sector by the Energy Policy and Planning Office. Yet, subsequent planning was done without major consultations, which resulted in some changes of rates and regulations. Notwithstanding the lack of public participation in the most recent phase of the program, renewable energy investors benefit from relatively high electricity purchasing rates, simplified procedures and secure contracts for a pre-specified period. As of December 2011, the operational capacity of renewable energy and the electricity generation under development combined to about 8,500 MW, which amounts to more than one-quarter of the installed electricity capacity (Tongsopit and Greacen, 2012).

*Table 13.1* Thailand's final energy consumption

| Final energy consumption | Quantity (ktoe) | | | Growth (%) | |
|---|---|---|---|---|---|
| | 2009 | 2010 | 2011ᴾ | 2010 | 2011ᴾ |
| Energy consumption | 66,698 | 70,247 | 70,562 | 5.3 | 0.4 |
| • Non-renewable energy | 54,243 | 56,829 | 56,807 | 4.8 | −0.04 |
| • Petroleum products | 31,661 | 32,096 | 33,078 | 1.4 | 3.1 |
| • Electricity | 11,512 | 12,724 | 12,681 | 10.4 | −0.3 |
| • Coal and its products | 7,493 | 8,240 | 6,563 | 10 | −20.4 |
| • Natural gas | 3,568 | 3,769 | 4,485 | 5.6 | 19 |
| • Renewable energy | 12,455 | 13,418 | 13,755 | 7.7 | 2.5 |
| • Commercial renewable energy* | 4,057 | 4,943 | 4,529 | 21.8 | −8.4 |
| • Traditional renewable energy** | 8,398 | 8,475 | 9,226 | 0.9 | 8.9 |

Source: Department of Alternative Energy Development and Efficiency (2011).

Notes
* Including solar, fuel wood, charcoal, paddy husk, bagasse, agricultural waste, garbage and biogas.
** Including fuel wood, charcoal, paddy husk and agricultural waste.
ᴾ Preliminary.

Yet it is still too early to conclude that the adder program has been a major driver of the success of renewable energy development. The sustainability of this process remains challenging, with political and institutional factors being the most decisive ones. One challenge is to overcome the lack of integration among departments under the MoE which have set different targets for renewable energy, leading to considerable confusion among the business community and the wider public as regards the future direction of the country's renewable energy policy.

Dissenting interests among the various electricity suppliers, particularly between the powerful former monopolist EGAT and the myriad of small and very small renewable energy providers, is another factor that has a bearing on future

*Table 13.2* VSPP and SPP renewable energy installed and with signed PPAs as of December 2011

| | Online (MW) | | Signed PPA (MW) | |
|---|---|---|---|---|
| | VSPP | SPP | VSPP | SPP |
| Solar | 3,304.37 | 381 | 102.97 | 8 |
| Biomass | 193.19 | 0 | 98.69 | 0 |
| Biogas | 2,029.13 | 387.5 | 349.52 | 375.2 |
| Municipal waste | 158.46 | 55 | 37.33 | 0 |
| Small hydro | 7.48 | 0 | 1.08 | 12.2 |
| Wind | 124.63 | 1,187.50 | 1.08 | 12.2 |
| Total | 5,547.26 | 2,011 | 589.96 | 395.4 |
| Grand total | 7,558 | – | 985.36 | – |

Source: Tongsopit and Greacen (2012).

developments in the sector. EGAT plays a dual role of electricity provider and buyer and sources its electricity not only from domestic generators, but also from neighbouring countries, such as Lao PDR, that can provide cheap electricity.

Another major challenge lies in the legal and regulatory framework. Thailand has yet to promulgate a comprehensive renewable energy act which would provide not only a stimulus, but also a clear legal mandate for all involved agencies, sectors and companies to align their actions towards raising the share of electricity generation from renewable sources. The existing legal vacuum shows a lack of political will to make renewable electricity provision from domestic sources a core element of Thailand's future energy strategy. This is underscored by the fledgling political and fiscal support of the adder program, which has created an atmosphere of uncertainty among domestic investors (cf. Tongsopit and Greacen, 2012).

The final challenge is associated with communication, information provision and public participation. The wider public in Thailand is neither informed about the energy sources behind the electricity that is consumed in the country nor aware of the emerging scarcity of future supplies. The lack of information and awareness triggers both overconsumption of electricity and limited interest in debates on the role of renewable energy. The public has also not been involved in debates on the acceptability of higher costs of electricity that will result from an enhanced share of renewable energy sources for electricity generation.

*Tax incentives for bioenergy promotion: the case of bioethanol*

The production and consumption of biofuels is supported by a combination of fiscal measures through various government bodies. The principal measure is the price mechanism through which the government keeps the price below that of the equivalent fossil fuel through a variety of tax exemptions as well as levy exemptions or reductions. As an additional measure, the Board of Investment (BOI) designed incentives, such as corporate tax exemptions. The MoF also contributed to the incentive package by revisions of the excise tax in 2007 and 2008 to promote vehicles that can run on gasoline–ethanol blends (i.e. gasohol) (Food and Agriculture Organization, 2009).

By the year 2021, the daily domestic consumption of bioethanol is set to reach nine million litres – from its current level of close to two million litres – with a projected annual increase of about 10 per cent according to the 10-Year Alternative Energy Development Plan (AEDP 2012–2021). In 2012, bioethanol production exceeded domestic consumption by about 40–50 million litres per day, making Thailand a major player in the global bioethanol market. Further increase in production will be mainly driven by rising foreign demand, with bioethanol exports already accounting for about 45 per cent of total ethanol production in 2012, compared to only 27 per cent in 2011 (Global Agricultural Information Network, 2012).

The government has made a number of strong arguments for further promotion of bioethanol, including national energy security, reduced dependency on fossil fuel imports, lower greenhouse gas emissions and the creation of new opportunities for strengthening local economies. Yet, a study by the Global Agricultural Information Network (2012) found that the Thai bioethanol market is dominated by large-scale, well-established sugar and cassava mills, with few opportunities for small- and medium-scale enterprise to compete. Another issue is the "biofuel versus food security" debate, since bioethanol produced from food crops has been identified as a major cause of rising food prices (Food and Agriculture Organization, 2010). In the European Union, the global food price hike in 2012 has recently caused a major rethinking among policy makers and a review of the goal of having a mandatory 10 per cent of road transport fuel stemming from biofuels by 2020 is currently under way (Reuters, 2012). Given these trends, the government may also need to readjust its high-flying goals for bioethanol production, to review its subsidy scheme for biofuels and to support renewable energy sources that do not compete with the food sector.

## Summary and concluding remarks

As Thailand emerges as an agro-industrialized country, a host of environmental problems has threatened its path towards sustainable development. The dynamics of environmental problems require appropriate policy reform. The move from the previous command-and-control policies towards the integration of economic incentive approaches under a more decentralized environmental governance regime has become an integral part of the political rhetoric of consecutive Thai governments. The newly proposed "Environmental Fiscal Act" aims at addressing environmental problems through behavioural changes induced by economic dis/incentive policy measures. Yet a number of barriers stemming from institutional and political factors have prolonged the process of having the act legalized. Among them are conflicts of interest and the notorious lack of coordination and collaboration among involved government agencies. Such barriers are also witnessed in the case of the implementation of public wastewater treatment charge which has long been introduced to solve the domestic wastewater problem. Implementation failures stem from communication failures between the public and the government, lack of public participation and the incapacity of local authorities to manage the wastewater treatment system.

Apart from reforming the environmental policy framework, Thai governments have also instigated significant reforms in the environment-related energy sector in recent years. A stronger liberalization of energy prices together with the promotion of renewable energy through various incentive measures is being implemented to achieve the main objectives of energy security and efficiency, while at the same time reducing adverse impacts on the environment. However, to maintain the government's political stability, the implementation of oil price subsidy removal, diesel in particular, has been prolonged. The adder program has played a pivotal role in promoting renewable energy in Thailand. However, the future of renewable energy will rely to a great degree on the improvement of various structural and institutional factors, including the integration of planning and implementation by the involved government agencies, the provision of legal basis, and the involvement of all relevant stakeholders in the process of policy design and planning. Investments in biofuel projects have shown upward trends, primarily as a result of a number of tax incentives provided by several government entities. However, the cooperation among various ministries and their line agencies needs a strong boost in order to come up with a comprehensive national plan that takes into account the balance between biofuel promotion, food security and sustainable development of the country.

In conclusion, Thailand appears to have arrived at a crossroads, and the Thai government faces a tough choice with far-reaching implications: it could either continue populist policies of providing cheap energy with disproportionate benefits for the better-off and adverse impacts on the state budget and the environment or it could decide on fiscally responsible measures for enhancing low carbon growth and instigate an integrative and inclusive reform package that will lead to enhanced energy efficiency, decentralized energy provision and far-sighted development of alternative energy sources without compromising the country's food security. Future generations will benefit if the Thai government opts for the latter.

## References

Asia-Pacific Economic Cooperation. 2005. Thailand's Energy Efficiency Revolving Fund: A Case Study. APEC Energy Working Group.

Credit Suisse. 2011. Thailand's Oil Price Subsidy: How Long Can it Last? Credit Suisse Economics Research, 2 March. Online, available at: www.credit-suisse.com/research-andanalytics.

Department of Alternative Energy Development and Efficiency. 2011. Thailand Energy Statistics 2011[(Preliminary)]. Ministry of Energy, Thailand.

Food and Agriculture Organization. 2009. Case Studies on Bioenergy Policy and Law: Options for Sustainability. FAO Legislative Study, the Development Law Service, FAO Legal Office, Rome, Italy.

Food and Agriculture Organization. 2010. Bioenergy and Food Security: The BEFS Analysis of Thailand. Bioenergy and Food Security Project, Rome, Italy.

Global Agricultural Information Network. 2012. Thailand Biofuels Annual. USDA Foreign Agricultural Service.

Greacen, Chuenchom Sangarasri and Greacen, Chris. 2012. Proposed Power Development Plan (PDP) 2012 and a Framework for Improving Accountability and Performance of Power Sector Planning. PalangThai, Thailand. Online, available at: www.palangthai.org/docs/PDP2012-Eng.pdf.

International Monetary Fund. 2012. Thailand: Staff Report for the 2012 Article IV Consultation. IMF, Washington, DC.

Kaosa-ard, Mingsarn, Laplante, Benoit, Rayanakorn, Kobkun and Waranyuwattana, Sakon. 2008. Thailand: Capacity Building for Pollution Taxation and Resource Mobilization for Environmental and Natural Resources Sectors – Phase II. Technical Assistant Consultant's Report, Asian Development Bank, Bangkok, Thailand.

Kititasnasorchai, Vipon and Tasneeyanond, Panat. 2000. Thai Environment Law. *Singapore Journal of International and Comparative Law*, 4, 1–35.

National Statistical Office of Thailand. various years. *Statistical Yearbook of Thailand*. National Statistical Office of Thailand, Bankok.

Pollution Control Department. 2012. Pollution Control Plan for 2012–2016 (in Thai). MoNRE, Thailand.

Rammont, Lalita and Amin, A.T.M. Nurul. 2010. Constraints in Using Economic Instruments in Developing Countries: Some Evidence from Thailand's Experience in Wastewater Management. *Habitat International*, 34, 28–37.

Reuters. 2012. Exclusive: EU to Limit Use of Crop-based Biofuels: Draft Law. Online, available at: http://uk.reuters.com/article/2012/09/10/us-eu-biofuels-idUKBRE8890SJ 20120910.

Simachaya, Wijarn. 2009. Wastewater Tariffs in Thailand. *Ocean & Coastal Management*, 52, 378–382.

Tongsopit, Sopitsuda and Greacen, Chris. 2012. Thailand's Renewable Energy Policy: FiTs and Opportunities for International Support. PalangThai, Thailand. Online, available at: www.palangthai.org/docs/ThailandFiTtongsopit&greacen.pdf.

World Bank and NESDB. 2008. Thailand Infrastructure Annual Report 2008. Online, available at: www-wds.worldbank.org/external/default/WDSContentServer/WDSP/IB/ 2012/01/04/000333037_20120104230930/Rendered/PDF/662200WP0p12440e0Clean 0Energy0all07.pdf.

World Bank and NESDB. 2011. Thailand: Clean Energy for Green Low-Carbon Growth. Online, available at: www-wds.worldbank.org/external/default/WDSContentServer/ WDSP/IB/2012/01/04/000333037_20120104230930/Rendered/PDF/662200WP0p124 40e0Clean0Energy0all07.pdf.

## Newspaper sources

*Bangkok Post*. 22 August 2012. Govt Raises Taxes on Alcohol, Cigarettes. Online, available at: www.bangkokpost.com/news/local/308728/govt-raises-taxes-onalcohol-cigarettes.

*Matichon* Online. 3 November 2010. ทศ.ค้านเก็บภาษีสิ่งแวดล้อม ชี้สอขัดกม.หน่วยงานในกระทรวง ทรัพย์ฯเกือบหมด (in Thai). Online, available at: www.matichon.co.th/news_detail.php?ne wsid=1288785281&grpid=&catid=19.

*Prachachart* Online. 4 November 2010. สุวิทย์ ขวางภาษีสิ่งแวดล้อม อ้างซ้ำซ้อนกองทุนสิ่งแวดล้อม นักวิชาการโต้ไม่ตอบโจทย์ (in Thai). Online, available at: www.prachachat.net/news_detail. php?newsid=1288847356.

*Thai Financial Post*. 14 June 2012. More to be Collected From Local Fuel Sales to Boost State Oil Fund. Online, available at: http://thaifinancialpost.com/2012/06/14/more-to-be-collected-from-local-fuel-sales-to-boost-state-oil-fund/.

*The Nation*. 18 August 2012. Diesel Subsidy to Hike Loss of Oil Fund. Online, available at: www.nationmultimedia.com/business/Diesel-subsidy-to-hike-loss-of-oil-fund-30188574. html.

*The Nation*. 24 July 2012. Diesel Excise Tax Waive Extended Again. Online, available at: www.nationmultimedia.com/business/Diesel-excise-tax-waive-extended-again-30186903.html.

# 14 Green fiscal policy strategies in response to climate change in Indonesia

*Budy P. Resosudarmo and Abdurohman*

## Introduction

As an archipelagic country, Indonesia is vulnerable to the impact of climate change. Prolonged droughts, increased frequency of extreme weather events, and heavy rainfall leading to floods, are a few examples of climate change impacts. In turn, these may have a harmful effect on agriculture, resulting in threats to food security (Jotzo *et al.* 2009). These potential negative environmental impacts, coupled with the fact that Indonesia is one of the three largest greenhouse gas emitters in the world, have prompted the Indonesian government to place a high priority on climate change issues and to aim to reduce emissions by 26 per cent from a business as usual scenario by 2020 and by 41 per cent if enough international support is available (Resosudarmo and Yusuf 2009).

It has been proposed that fiscal policy could play a key role in achieving this emission reduction target; not only as a means by which potential carbon tax systems or other relevant policy schemes could be introduced, but also to eliminate environmentally unsustainable biases such as energy subsidies, low forestry rents, and improper profit sharing schemes in the mining sector which have distorted price signals, leading to overconsumption and exploitation. The aim of this chapter is first to review the existing environmental management fiscal policy regime in Indonesia, from both the revenue and expenditure perspectives. With regard to expenditure, broad structures and trends of budget allocation for green activities will be examined. Concerning revenue, various green taxes and levies that have been or once were implemented in Indonesia will be outlined. Second, this chapter will discuss the structure of $CO_2$ emissions and the challenges involved in their reduction. Third, this chapter reviews recent fiscal policies in response to Indonesia's commitment to reduce its emissions by 2020. Final remarks concern the opportunities and challenges faced by Indonesia in achieving its emission reduction target.

## Key green fiscal policies in Indonesia

This section reviews some of the green fiscal policies that various agencies in the country have attempted to implement.

## Ministry of the Environment

In 1978 President Suharto demonstrated Indonesia's concern for environmental issues when he established Indonesia's first Ministry of the Environment.[1] Since then, a fiscal budget has been allocated to this Ministry to design regulations and programmes to manage the country's environment. The first environmental law to be enacted was Law No. 4/1982 on Basic Provisions of Environmental Management (Koesnadi 1989). Since then, various laws, regulations, and programmes related to the environment have been developed; not only by the Ministry of the Environment, but also by other ministerial agencies.

In 2010, the budget for the Ministry of the Environment was approximately Rp416 billion[2] or approximately 0.06 per cent of central government spending. This allocation, though very small, is certainly one of Indonesia's key green fiscal policies enabling the Ministry to develop several, though not all, of its environmental management programmes. When environmental budgets of other ministries are included, the total central government environmental budget, in 2010, was approximately 1.1 per cent of the total central government budget (Table 14.1). This situation reflects the still relatively low importance of environmental issues in Indonesia and the relatively little influence the Ministry of the Environment has within the cabinet.

Examples of programmes that have been developed by the Ministry of the Environment are as follows. The first serious attempt to control pollution was made in 1989 when the Ministry of the Environment, by means of Government Regulation No. 29/1986, started to require all firms and development programmes to submit a report on the expected impact of their activities on the environment, their proposed actions to avoid or minimise this impact, and what environmental impact monitoring activities they would undertake. This report – called an environmental impact assessment (*Analisa Mengenai Dampak Lingkungan* or AMDAL) – is assessed by whichever government agency has the authority to permit the activity. If the project is expected to have a serious

*Table 14.1* Government spending on the environment

| Year | Spending on environment (Rp billion) | Share of total central government spending (%) |
|------|--------------------------------------|------------------------------------------------|
| 2005 | 1,334 | 0.37 |
| 2006 | 2,665 | 0.61 |
| 2007 | 4,953 | 0.98 |
| 2008 | 5,315 | 0.77 |
| 2009[1] | 6,684 | 0.96 |
| 2010[2] | 7,753 | 1.11 |

Source: Ministry of Finance's Database 2010, raw data.

Notes
1 Revised Budget of 2009 (APBN-P).
2 Budget 2010.

environmental impact, the report needs to be assessed by an inter-sectoral team involving the Ministry of the Environment.

The Clean River Programme, also known as PROKASIH (*Program Kali Bersih*), was started in 1989 in response to the declining quality of rivers in Jakarta. It is based on a consensual approach to control wastewater effluents by targeting selected industrial polluters in priority river basins. The goal is to improve river quality by setting a specified standard for river water based on some parameters – usually biochemical oxygen demand (BOD) and total suspended solids (TSS) – established by the government (Resosudarmo *et al.* 1997; ADB 1997). This programme also attempts to strengthen human and institutional capabilities regarding the management of river water quality and of riverbanks (Resosudarmo 2003).

In June 1990, through Presidential Decree No. 23, a national environmental impact management agency (*Badan Pengendalian Dampak Lingkungan* or BAPEDAL) was established (Hardjono 1994). The main task of this agency is to implement various programmes developed by the Ministry of the Environment.

The PROPER programme was established in June 1995 as an alternative environmental policy developed by the Ministry of the Environment. PROPER is basically an environmental rating programme that acts as a simple form of environmental certification. Its main objectives are to increase compliance with environmental regulations, promote adoption of clean technologies, create incentives for polluters to strengthen their in-house environmental capabilities, and prepare companies in Indonesia for International Standards Organization (ISO) 14001 certification (Wheeler and Afsah 1996; ADB 1997).

Implementation of this programme was conducted by BAPEDAL. Each year BAPEDAL evaluated companies participating in this programme with regard to their environmental abatement activities. Based on their success in reducing waste discharge, mainly water effluent – hence the programme was also called PROPER PROKASIH – these companies were given one of five different colour scores (ADB 1997). The results were published in newspapers and other media to make them known publicly.

In 2002, since BAPEDAL was merged into the Ministry of the Environment, the Ministry of the Environment took over and developed a new PROPER programme with the more comprehensive goal of multi-media management, aiming not only to control water, air, and toxic waste discharge, but also to be involved in environmental impact assessment (AMDAL). It is also equipped with new regulations such as Government Regulation No. 27/1999 on environmental impact assessment (AMDAL), Regulation No. 18/1999 on toxic waste management, Regulation No. 82/2001 on air and water pollution, Ministerial Decree No. 113/2003 on waste water discharge, and Ministerial Decree No. 129/2003 on emission (Ardiputra 2004).

Other than PROPER, there are few monitoring activities conducted for hazardous waste disposal. Only a small number of industries dispose of their hazardous waste at the existing treatment facilities and the number of hazardous waste treatment plants and facilities available in a treatment plant are limited. In fact, the Cileungsi Hazardous Waste Treatment Plant near Jakarta (operational in 1994) is the only treatment facility capable of processing this waste and providing a secure storage area and a lined landfill for the disposal of stabilised and low level toxins (World Bank 2003).

## Central government environmental expenditure

In Indonesia, government spending is conducted by both central and regional governments. The central government spending or budget is the amount that is spent by the central government, typically through sectoral ministries, and is generally approximately 70 per cent of the total government budget. The rest is spent by regional governments and is called regional spending.

As mentioned before, besides funds allocated to the Ministry of the Environment, the Indonesian central government does provide funding for other programmes related to the environment. This could be funding for reforestation allocated to the Ministry of Forestry, or for building dams allocated to the Ministry of Public Works. The total central government budget for environmental activities, excluding the regional government budget, in the last several years can be seen in Table 14.1. This total is much higher than what is allocated through the Ministry of the Environment, as it only controls 5 per cent of this environmental budget.

It can be seen from Table 14.1 that, though relatively small, the amount – and also its share – of the government budget for environmentally related programmes has increased. This is an indication that the Indonesian government is becoming more concerned about the condition of the country's environment, though it is still not a top government priority. By comparison, the education budget is approximately 20 percent of the total government budget.

Table 14.2 represents the distribution of central government environmental spending. The four categories are waste management, pollution mitigation, natural resource conservation, and land use management. Most funding is spent

*Table 14.2* Percentage distribution of government environmental spending

|  | 2008 | 2009 |
|---|---|---|
| 1 Waste management | 8.4 | 7.5 |
| 2 Pollution mitigation | 3.5 | 2.5 |
| 3 Natural resource conservation |  |  |
| • Conservation | 6.9 | 5.9 |
| • Rehabilitation | 5.9 | 9.5 |
| • Capacity development | 6.8 | 4.7 |
| • Flood control and coastal protection | 37.5 | 36.9 |
| 4 Land use management | 28.1 | 30.7 |
| 5 Other environment | 2.8 | 2.3 |
|  | 100 | 100 |

Source: Ministry of Finance's Database (2010).

on natural resource conservation, in particular on flood control and coastal protection, with the majority of the funding going to flood control. Hence, it can be observed that the two main environmental priorities are controlling floods and managing land use. This is probably a good strategy since these issues are problematic and at the same time affect many people, particularly in Java.

### Decentralised environmental funds

Regional governments also spend some of their budget on environmental programmes. As with the central government, this spending can be for building dams, improving waste management or water treatment works, among other things. The central government transfers certain funds to regional governments designated to be spent only on specific programmes, namely special allocation funds (*Dana Alokasi Khusus* or DAK). Up till 2012, there have been two types of DAK that relate to the environment: environmental DAK and forestry DAK (Table 14.3). Environmental DAK supports activities like water treatment, air pollution control, and waste management. This DAK is specifically allocated to regions that face significant environmental problems and have been left behind. Distribution among regions in Indonesia can therefore be on the basis of environmental problems as well as whether a region lags behind. Java-Bali, for instance, has received large amounts of environmental DAK funding, reflecting its significant need for clean water, waste, and air pollution management. Sumatra and Eastern Indonesia have received relatively large amounts due to their high number of regions that lag behind.

Forestry DAK supports reforestation in particular. Distribution of these funds among regions in Indonesia reflects not only the need but also the ability of regional governments to reforest their forest areas. Kalimantan, for example, should have a high DAK since the rate of deforestation in the area is relatively faster than in other areas. However, it does not receive a large amount of DAK due to the fact that revenues of regional governments in Kalimantan are

*Table 14.3* Special allocation fund (DAK) for the environment (Rp billion)

| Region | Environment | | | | Forestry | |
|---|---|---|---|---|---|---|
| | *2006* | *2007* | *2008* | *2009* | *2008* | *2009* |
| Sumatra | 29.7 | 101.9 | 101.9 | 96.0 | 29.8 | 39.4 |
| Java and Bali | 30.4 | 78.5 | 78.5 | 70.6 | 8.6 | 15.2 |
| Kalimantan | 12.4 | 40.1 | 40.1 | 40.6 | 11.5 | 3.9 |
| Sulawesi | 16.5 | 57.4 | 57.4 | 62.0 | 23.3 | 18.9 |
| Eastern Indonesia | 23.8 | 73.8 | 73.8 | 82.3 | 26.8 | 22.7 |
| Total | 112.9 | 351.6 | 351.6 | 351.6 | 100.0 | 100.0 |

Source: Ministry of Finance's Database (2010).

Note
Eastern Indonesia=Nusa Tenggara, Maluku, and Papua.

relatively high. In general, the total amount of DAK for the environment and forestry is relatively less than 1 per cent of total regional government budgets.

Regional governments also develop and fund their own environmental programmes. Table 14.4 represents the amount of funding spent on these activities. In general the amount is approximately 3–6 per cent of their development budget, but still less than 1 per cent of their total budget. In Java-Bali, a significant portion of this amount is for water management, such as flood control and water distribution. In Sumatra and Kalimantan, it is for activities related to forest management, including reforestation.

In general, it can be seen that budget allocations to environmental programmes by regional governments are increasing. However the amounts allocated are still relatively small. Environmental management is not yet a top priority for regional governments.

### Taxes and charges

Several user charges have been levied throughout Indonesia. These are specifically related to municipal services such as drinking water, wastewater treatment, solid waste collection and disposal, and to road/transportation services (ADB 1997; Rock 2000). Several examples related to industrial pollutants are as follows.

Regarding water management, in most provinces a surface or raw water use licence is required for withdrawing surface water, with the licence fee starting at US$43 in 1996. For six main classes of customers, a volume-based fee is added, ranging from zero for households to Rp150/m$^3$ for industrial users (ADB 1997; Rock 2000). These prices are considered low, particularly for industrial users, and hence it is expected that these prices do not really affect personal and industrial water use behaviour.

Furthermore, there are around 300 PDAMs (state-owned municipal water companies) providing access to treated pipe water for around 36 per cent of the urban population throughout the country, with fees averaging from Rp350–2,000/m$^3$. These prices are considered too low, and so many of these PDAMs are in financial difficulty. In 2004, the government enacted a new law allowing private enterprises to manage the water supply in various regions across the country (PERPAMSI 2004).

Regarding sewage disposal, in 1989 the Ministry of Public Works decided to apply a sewage disposal levy. This levy covers five categories of users and in 1997 was determined according to the floor space of a building, office, or house. The rate in 1990 was from Rp28/m$^2$ for households to Rp182/m$^2$ for commercial high-rise buildings (ADB 1997; Rock 2000). Local governments are responsible for the management of sewage disposal collection, which is generally conducted either by contractors or residents themselves. Only a miniscule portion of the solid waste is recycled, and this by the informal private sector (e.g. scavengers) in many large cities in Indonesia (World Bank 2003).

Table 14.4 Regional government spending on the environment

| Region | Environment Spending (Rp billion) | | | | | Share to total development expenditure (%) | | | | |
|---|---|---|---|---|---|---|---|---|---|---|
| | 1995 | 1997 | 1999 | 2001 | 2002 | 1995 | 1997 | 1999 | 2001 | 2002 |
| Sumatera | 41.8 | 72.5 | 132.7 | 180.3 | 327.7 | 2.28 | 2.93 | 4.00 | 2.05 | 2.35 |
| Jawa and Bali | 132.7 | 177.7 | 346.1 | 494.0 | 692.4 | 3.28 | 3.43 | 6.50 | 4.15 | 4.57 |
| Kalimantan | 31.9 | 28.7 | 61.0 | 130.9 | 467.6 | 3.49 | 2.52 | 3.95 | 2.81 | 7.89 |
| Sulawesi | 31.5 | 40.3 | 83.5 | 50.7 | 79.3 | 4.25 | 3.91 | 7.32 | 2.66 | 3.09 |
| Eastern Indonesia | 22.3 | 31.7 | 75.7 | 72.8 | 112.7 | 2.96 | 3.89 | 6.08 | 2.63 | 2.59 |

Source: Ministry of Finance's Database (2010).

Note
Eastern Indonesia=Nusa Tenggara, Maluku, and Papua.

The user charge was less effective for solid wastes than it was for water management, mainly because the agency that managed the solid waste did not collect the fees. Solid waste management was funded through a Presidential Instruction Decree (Inpres) by provincial governments, while local/district financial offices or *Dinas Pendapatan Daerah* (Dispenda) collected the fees (Rock 2000) and there was a failure of collaboration between the two bodies. Due to the lack of an effective user charge system, solid waste collection in urban areas ranges from 40 to 85 per cent of total wastes, including the share attributable to scavengers (ADB 1997; Rock 2000). In several areas though, due to financial strain, the government privatised some collection services and most of these were considered to be more successful (Rock 2000).

Since the 1980s, a fuel tax of as much as 5 per cent has been in place in Indonesia in addition to a 10 per cent sales tax. The main goal, though, is to raise the revenue of local governments. A total of 10 per cent of the total fuel tax revenue goes to the central government, and the rest is distributed to provincial, district, and municipal governments. Since 1997, however, the government has subsidised domestic fuel prices by much more than 5 per cent, so this fuel tax has not been effective in controlling its use.

Table 14.5 shows fuel pricing and its tax component for 2009. The Ministry of Finance (MoF) calculated the benchmark price representing the costs incurred by the National Oil Company (PERTAMINA) in producing and distributing fuel. In doing so, the MoF bases the benchmark price on the MOPS (Mid Oil Platts Singapore) plus alpha magnitude (parameter of distribution cost). The pre-tax price is the difference between the benchmark price and the subsidy. Value-added tax (VAT) and local fuel tax amounts are based on this pre-tax price. The retail price is the sum of the pre-tax price, VAT, and local fuel tax. It can be seen that, due to the subsidy, the retail price, even when including taxes, is lower than the benchmark price, meaning that taxes are not effective. It is also clear that, were there no subsidy, under the existing tax arrangement the price should be as shown in the market price column.

Table 14.6 shows the size of the fuel and electricity subsidies over several years. It can be seen that, though declining, the total subsidy allocated to fuel and electricity has been relatively high, since prices of electricity and domestic fuel are controlled by the government and do not relate to world prices. The level of government subsidisation has fluctuated over the years depending on the price of world crude oil.[3] These subsidies have distorted the economy and are bad for the environment. It reduces the incentive to develop and use renewable energy sources and encourages higher levels of emission.

The government has imposed various taxes and fees on the extraction of natural resources. These are related to oil, gas, geothermal, mineral, fish, and forest extraction activities. Each has a different tax and fee system. Table 14.7 illustrates the size of these taxes and fees related to natural resource extraction activities. It can be seen that revenues from oil and gas have been the most significant of the natural resource revenues. Mining has become more significant since the mid 2000s. Forestry revenues have declined, indicating the declining size of the forest area in Indonesia.

Table 14.5 Fuel pricing policy and tax component

| Fuel type | Retail price (Rp/ltr) | Pre-tax price (Rp/ltr) | Tax | | Benchmark price (Rp/ltr)[1] | Subsidy (Rp/ltr)[2] | Market price (Rp/ltr)[3] |
|---|---|---|---|---|---|---|---|
| | | | VAT, 10% (Rp/ltr) | Fuel tax (PBBKB), 5% (Rp/ltr) | | | |
| Premium | 4,500 | 3,913 | 391 | 196 | 4,711 | 798 | 5,417 |
| Kerosene | 2,500 | 2,273 | 227 | – | 4,925 | 2,652 | 5,663 |
| Diesel oil | 4,500 | 3,913 | 391 | 196 | 4,854 | 941 | 5,582 |

Source: author calculation based on Ministry of Finance's Database (2010).

Notes

All figures in the table are based on the 2009 government actual budget.

1 The benchmark price is the MOPS plus alpha magnitude (parameter of distribution cost).
2 The subsidy is the difference between the benchmark and pre-tax prices.
3 The market price is the benchmark price plus tax (10% VAT and 5% fuel tax). This is the price that should have prevailed without the subsidy.

*Table 14.6* Fuel and electricity subsidy 1998/1999 to 2009 (Rp billion)

| Year | Fuel | Electricity | Total for energy | Share of total central government spending (%) |
|------|------|-------------|------------------|-----------------------------------------------|
| 1998/1999 | 28.6 | 1.9 | 30.5 | 21.3 |
| 1999/2000 | 40.9 | 4.6 | 45.5 | 22.5 |
| 2000 | 53.8 | 3.9 | 57.7 | 30.6 |
| 2001 | 68.4 | 4.6 | 73.0 | 28.0 |
| 2002 | 31.2 | 4.1 | 35.3 | 15.8 |
| 2003 | 30.0 | 3.8 | 33.8 | 13.2 |
| 2004 | 69.0 | 2.3 | 71.3 | 24.0 |
| 2005 | 95.6 | 8.9 | 104.5 | 28.9 |
| 2006 | 64.2 | 30.4 | 94.6 | 21.5 |
| 2007 | 83.8 | 33.1 | 116.9 | 23.2 |
| 2008 | 139.1 | 83.9 | 223.0 | 32.2 |
| 2009 | 45.0 | 49.5 | 94.6 | 14.7 |

Source: Ministry of Finance's Database (2010).

Notes
Up to April 2000, the fiscal year is from April in the current year to March the following year. Since April 2000, the fiscal year is from January to December (calendar year).

In total, the revenue from natural resource extraction activities is significant for the government, amounting to around 20–30 per cent of its total revenue. The government should impose a higher rate of taxes and fees to reduce the rate of resource extraction in the country.

However, as long as illegal extraction of natural resources is rampant, a higher rate would mean a lower level of compliance and would stimulate even more illegal activity, leading to an even higher rate of resource extraction and less income for the government. Another important consideration is that a considerable number of Indonesians are employed in natural resource extraction activities. A reduction in activities would mean fewer available jobs.

## Greenhouse gas emission in Indonesia

During the 1970s through 1990s, the Indonesian economy, measured by its gross domestic product (GDP), grew at an average annual rate of above 7 per cent, which was relatively fast compared to many other countries. The country's economy had also been transformed from an agriculture-dominated economy to a more industry-dominated one (Hill 2000; Resosudarmo and Kuncoro 2006; Resosudarmo and Vidyattama 2007). These factors contributed to an increased amount of $CO_2$ emitted by fossil fuel combustion.[4] Consequently, there are three important facts related to Indonesia's $CO_2$ emission from fossil fuel combustion.

First, Indonesia is one of the highest $CO_2$ emitter countries in the world (IEA 2006). Since 2004, Indonesia is among the top 25 $CO_2$ emitters – ranked sixteenth when counting the EU as one country – due to fossil fuel combustion (PEACE 2007; Resosudarmo *et al.* 2011a). The exact ranking is somewhat sensitive; i.e. it

Table 14.7 Government revenue from natural resources

| Items | 2002 | 2004 | 2006 | 2008 | 2009 |
|---|---|---|---|---|---|
| Non-tax revenue (Rp billion) | 88,440 | 122,546 | 226,950 | 320,605 | 227,174 |
| Natural resources | 64,755 | 91,543 | 167,474 | 224,463 | 138,959 |
| Oil and gas | 60,011 | 85,259 | 158,086 | 211,617 | 125,752 |
| Oil | 47,686 | 63,060 | 125,145 | 169,022 | 90,056 |
| Gas | 12,325 | 22,199 | 32,941 | 42,595 | 35,696 |
| Non-oil and gas | 4,744 | 6,284 | 9,388 | 12,846 | 13,207 |
| Mining and quarrying | 1,457 | 2,549 | 6,781 | 9,511 | 10,369 |
| Charges | 103 | 196 | 171 | 103 | 146 |
| Royalty | 1,354 | 2,352 | 6,610 | 9,408 | 10,223 |
| Forestry | 3,130 | 3,412 | 2,409 | 2,316 | 2,345 |
| Iuran Hak Pengusahaan Hutan (IHPH) | 28 | 90 | 112 | 68 | 74 |
| Provisi Sumber Daya Hutan (PSDH) | 800 | 907 | 575 | 620 | 677 |
| Reforestation fund | 2,302 | 2,415 | 1,691 | 1,627 | 1,425 |
| Other | – | – | 31 | – | 170 |
| Fishery | 157 | 324 | 197 | 78 | 92 |
| Geothermal | – | – | | 941 | 400 |
| Income tax revenue from oil and gas (Rp billion) | 17,469 | 22,947 | 43,190 | 55,987 | 50,044 |
| Share of total non-tax revenue (%) | | | | | |
| Natural resources | 73.2 | 74.7 | 73.8 | 70.0 | 61.2 |
| Oil and gas | 67.9 | 69.6 | 69.7 | 66.0 | 55.4 |
| Non-oil and gas | 5.4 | 5.1 | 4.1 | 4.0 | 5.8 |
| Mining and quarrying | 1.6 | 2.1 | 3.0 | 3.0 | 4.6 |
| Forestry | 3.5 | 2.8 | 1.1 | 0.7 | 1.0 |
| Fishery | 0.2 | 0.3 | 0.1 | 0.0 | 0.0 |
| Other | 26.8 | 25.3 | 26.2 | 30.0 | 38.8 |
| Share of total government revenue (%) | | | | | |
| Natural resources (without oil and gas income tax) | 21.7 | 22.7 | 26.3 | 22.9 | 16.0 |
| Natural resources (with oil and gas income tax) | 27.5 | 28.4 | 33.1 | 28.6 | 21.8 |

Source: Ministry of Finance's Database (2010).

Notes
IHPH and PSDH are typically translated as forest royalties.

could easily change, as there are many countries that have an only slightly lower level of emission than Indonesia. If $CO_2$ emission due to deforestation and land use change is included, since it is approximately twice that from fossil fuel combustion, Indonesia then becomes among the top three to five emitters (PEACE 2007; Resosudarmo *et al*. 2011a). It is important to note, however, that the reliability of $CO_2$ emission data from deforestation and land use change is questionable.

Second, in terms of $CO_2$ emission per capita from fossil fuel combustion, Indonesia's emission per capita is still low (approximately two tons per person in mid 2000s) in comparison with other countries; however, it is growing relatively fast (approximately 6 per cent annually), exceeding that of China and India between 1994 and 2004 (IEA 2006; Resosudarmo *et al*. 2011a, 2011b).

Third, in the long run, $CO_2$ emission from fossil fuel combustion will most likely be much more important than that caused by deforestation and land use change, although the reverse is true at present. In the future, if Indonesia does nothing, there should be a tendency for lower deforestation emissions as the rate of deforestation will be slower due to a decrease in available forest areas; whereas energy use and fossil fuel emissions will keep growing as GDP grows unless mitigating actions are taken (IEA 2007; Resosudarmo *et al*. 2011b).

As a consequence of the three facts related to $CO_2$ emission from fossil fuel combustion mentioned above, in order to maintain a low carbon growth in Indonesia, a two-pronged strategy must be implemented: reduce deforestation and lower the trajectory of fossil fuel emission. Immediately effective action to reduce deforestation, such as through the Reducing Emissions from Deforestation and Forest Degradation (REDD) programme,[5] could significantly reduce the country's level of $CO_2$ emission and preserve the forests. With regard to fossil fuel combustion, it is crucial to bear in mind that investments in the energy sector have long lifetimes, often more than 35 years. Investments made now are locked in a carbon emission trajectory for a long time (IEA 2003).

The literature typically identifies the drivers of deforestation in Indonesia as: (1) land use change due to plantation and mining activities as well as population growth; (2) intensive illegal logging due to high domestic and international demand for timber, weak law enforcement, and the existence of poverty and social issues among forest communities; (3) forest fires due to land clearing activities mostly for plantations and due to natural forces such as long dry seasons and peat forest; (4) limited practice of selective cutting due to a lack of enforcement as well as its being difficult and costly to implement; (5) limited implementation of forest plantation due to ease of access to natural forests (Sunderlin and Resosudarmo 1996; Barber *et al*. 2002). The debate has been over who causes how much deforestation. Exact data to resolve this debate is limited.

Another important fact related to Indonesia's $CO_2$ emission from fossil fuel combustion is that its intensity, though has been declining, is still positively growing. Meanwhile, in many countries, including developing countries, it has been negative in the last decade (Resosudarmo *et al*. 2011b). Positive growth in emission intensity means that emission growth is faster than economic growth. Decomposition analysis of Indonesia's emission intensity into energy and carbon intensities shows that both increased in the last decade, and the growth rates of carbon intensity were always higher than those of energy intensity. This indicates that increasing the carbon intensity of energy supply is the more important driver than energy intensity for the increase in emission intensity in Indonesia (IEA 2007; Resosudarmo *et al.* 2011b).

To understand why carbon intensity in Indonesia has increased significantly, we need to observe the level of emissions for the combustion of each type of fossil fuel (or energy source) used in the country. Data from the International Energy Agency shows that the share of $CO_2$ emission from oil combustion has declined from contributing 85 per cent of total $CO_2$ emissions in the mid 1980s to 53 per cent in the mid 2000s (IEA 2007; Resosudarmo *et al.* 2011b). Meanwhile, during the same period of time, the contributions of gas and coal combustions increased from 14 to 21 per cent and from 1 to 26 per cent, respectively (IEA 2007; Resosudarmo *et al.* 2011b).

The broad trend in Indonesia's energy system is that, although oil is still the main contributor of $CO_2$ emissions, its share has been decreasing. Most oil products are used in the transportation sector, with oil use in electricity generation declining strongly. The share of gas roughly remains constant, with expanding gas production directed to a large degree at exports. In the meantime, the share of coal has been increasing and takes up the role of oil in the stationary energy sector; i.e. in electric power generators.

## Fiscal policy response to climate change in Indonesia

At the 2009 G20 meeting in Pittsburgh, President Susilo Bambang Yudhoyono announced that Indonesia had committed to a national climate change action plan that will reduce its emissions by 26 per cent by 2020 from BAU (Business As Usual), and that with international support Indonesia 'could reduce emissions by as much as 41 per cent' (Resosudarmo and Yusuf 2009). Much of the potential for reduction (more than 80 per cent) relates to forestry, peat-land, and agriculture, where Indonesia makes its greatest contribution to global warming. This commitment to reduce greenhouse gas emissions means Indonesia now engages with climate change policy at a new level, so that it is now part of the national policy framework, including the national economic policy framework. In 2011, Presidential Decree No. 61/2011 on the National Action Plan for Greenhouse Gases Reduction (*Rencana Aksi Nasional Penurunan Emisi Gas Rumah Kaca* or RAN-GRK) was enacted.

Within the national economic policy framework, a sound set of economic policies is the key to mitigating $CO_2$ emission; in particular appropriate pricing of $CO_2$ and fiscal policy plays a central role in shaping Indonesia's response to climate change challenges. Central to Indonesia's climate change policy is the implementation of the REDD+ programme. The catalyst to implement this programme is a May 2010 commitment of $1 billion by the government of Norway to support the implementation of REDD+ in Indonesia. The agreement makes phased transfers of the funds conditional on the implementation of pilot schemes and a deforestation moratorium, the establishment of a REDD+ strategy and institutions, and measured achievements in reducing emissions. The president hence appointed the President's Delivery Unit for Development Monitoring and Oversight (*Unit Kerja Presiden Bidang Pengawasan dan Pengendalian Pembangunan* or UKP4) to manage initial implementation, and signed a presidential decree (No. 25/2011) establishing a REDD+ Task Force. The task force has established ten working groups to develop the policy and institutional infrastructure for REDD+. A REDD+ National Strategy, released in June 2012, proposes the establishment of several implementing institutions: a REDD+ agency (to govern and coordinate all REDD+ activities); a REDD+ funding instrument (to manage and facilitate the distribution of funds); and a REDD+ measurement, reporting, and verification institution (Indonesia REDD+ Task Force 2012). There are currently over 45 demonstration activities, about half of them in Kalimantan (Forest Climate Center 2012).

There are numerous challenges to the successful implementation of REDD+ in Indonesia. Negotiations with international partners are proceeding slowly, and the prospects for a functioning international market in permits from emission reductions from deforestation within the next few years appear slim. Achieving broad institutional backing for REDD+ at both the central and local government levels has also been difficult. There is a risk that the REDD+ agency will be relatively weak. Challenges on the ground are also considerable. Ensuring that payments actually reach those whose actions result in reduced emissions is difficult, and there are serious difficulties with land tenure, forest monitoring, and emissions measurement. Conflicts within communities, corruption, and weak law enforcement capacity further complicate REDD+ implementation. Profits from alternative land uses such as oil palm and mining are likely to remain higher than any payments for avoided deforestation for the foreseeable future, meaning that private incentives will continue to favour forest clearing (Burke and Resosudarmo 2012). Appropriate fiscal policies are needed to support the implementation the REDD+.

Within the Ministry of Finance, the broad plan regarding fiscal policies related to climate change is as follows (Ministry of Finance 2009; BAPPENAS 2010):

- Energy sector:

  a   work towards the implementation of a carbon tax/levy on fossil fuel combustion, in parallel with removal over time of the energy subsidy;

  b   introduce complementary measures to incentivise energy efficiency and deployment of low-emissions technology, exemplified by a specific geothermal policy strategy;

  c   access to international carbon markets.

- Land-use change and forestry sector:

  a   support and incentivise carbon abatement measures by regional governments through the intergovernmental fiscal transfer system, working towards the creation of a Regional Incentive Mechanism (RIM) for climate change;

  b   work with the appropriate ministries to bring existing fiscal policy set-
      tings into line with carbon reduction objectives.

- International carbon finance:

  a   support the creation of new, broad-based carbon market mechanisms
      like sectoral targets and crediting;
  b   support new and additional sources of international public financing;
  c   ensure adequate returns for Indonesia's emissions reductions.

- Institutional development:

  a   strengthen the capacity for climate policy analysis at the Ministry of
      Finance;
  b   support policy coordination across government especially among the
      economic ministries;
  c   advocate a review of the broader regulatory framework that relates to
      climate change policy.

In implementing the above strategy, the Ministry of Finance is expected to col-
laborate with other relevant ministries. For example, in the energy sector, the
MoF together with other economic ministries and the Ministry of Energy and
Mineral Resources (the MEMR) has launched a Roadmap of Fuel and Electricity
Subsidy Reduction. The fuel subsidy is expected to be completely eliminated in
2014, and transformed into a direct subsidy for the neediest group. Meanwhile,
the electricity subsidy will gradually move towards only targeting those consuming between 450 VA and 1,300 VA. In addition, a number of policies have been
launched to accelerate renewable energy development in Indonesia, in particular
geothermal. There are several fiscal incentives, one of which is a plan to allocate
a revolving fund in the 2011 budget to help investors in the exploration stage. In
institutional development, a working group team has been set up within the MoF
which has been in communication with other ministries to discuss policies on
climate change related issues.

There are certainly challenges in implementing the strategies mentioned
above. For example, in the energy sector, the more important challenges are as
follows. First, so far the removal of the fuel subsidy is not a popular policy. Any
announcement of a possible increase in the price of fuel creates street riots and
social unrest. It is extremely challenging for the government to be able to
develop a carbon tax and the elimination of the fuel subsidy by 2014. Second,
currently there is a crash programme to develop coal power electric plants with
as much as an 11,000 MW capacity; most of them will be in Java. Whereas the
direction of climate change policy is to reduce the use of coal, particularly for
electric power generation, this programme encourages Indonesia to use even
more coal for its electricity. Since this is a crash programme, it is doubtful that
many of these power plants will rely on critical/high coal power plant technology. In addition, since the amount of coal needed for these power plants will be
high, they will rely on low grade coal supplies.

Up till the end of 2012, the broad plan mentioned above reminds a plan, the MoF has not made any significant progress.

## Final remarks

The expected devastating impacts of climate change have revived the discussion regarding the importance of green fiscal policy as a mechanism to tackle the environmental externality problem. In this case it is the externality of causing too much $CO_2$ to be emitted that ultimately has induced worldwide climate change. In Indonesia, as one of the more significant emitters of $CO_2$, the discussion of green fiscal policy is also taking place. This chapter has reviewed the existing regime of fiscal policy in Indonesia related to environmental management both from the revenue and expenditure aspects. It has presented the structure of $CO_2$ emissions and challenges for Indonesia in reducing its emissions, as well as recent fiscal policies in response to Indonesia's commitment to reduce its emissions by 2020.

Several observations can be drawn from this review. First, there have been green components in Indonesia fiscal policy for a long time. Second, a significant milestone in supporting the development of green fiscal policy was the establishment of a ministerial level agency, the Ministry of the Environment, at the end of the 1970s. However, the budget allocated to this Ministry for developing various policies and programmes to protect the environment has been modest.

Third, a much greater environmental budget is available to other ministries, such as the Ministry of Public Works for flood control and the Ministry of Forestry for land use management. Total central government expenditure to support environmental programmes has increased over time, though the amount, at least up until 2010, has been very small.

Fourth, within the budget for environmental programmes, the top priorities are for flood control and land use management. Fifth, regional governments also allocate some of their budget towards environmental activities but, again, the amount is relatively small.

Sixth, user charges and fees for natural resource use and extraction have also been developed. However, user charges are generally ineffective in controlling and stimulating a more efficient use of resources. The situation is worse in the case of fuel and electricity, because of government subsidies. Various fees related to natural resources have provided significant revenue for the government. The government's intention is to raise revenue, however, rather than to effectively control the rate of extraction.

Seventh, the main sources of Indonesia's $CO_2$ emission have been identified, i.e. deforestation and the use of coal in the energy sector, and so recent fiscal policies have also been developed to mitigate this emission. The main fiscal policies are to support the development of the REDD+ programme and to impose a carbon tax together with the elimination of the fuel and electricity subsidy, to create an energy mix favouring geothermal and renewable resources. Nevertheless the challenges with regard to implementing these policies are not trivial. Among others are, first, the perennial difficulty in finding disincentives to converting natural forests;

second, an increase in energy prices typically induces social unrest; and third the ongoing crash programme to develop coal-based electric power plants.

Finally, we conclude that green fiscal policies in Indonesia have progressed. However, it remains uncertain as to how effective they are in solving Indonesia's environmental issues. Certainly a more vigorous approach is needed to be able to protect Indonesia's environment and to cope with the new challenges of controlling $CO_2$ emission in the era of climate change.

## Notes

1 Initially this agency was the State Ministry of Development Supervision and the Environment. In 1983, it was renamed the State Ministry of Population and the Environment. In 1993, it became the State Ministry of the Environment. In this chapter, we will refer to all of them as the Ministry of the Environment.
2 It is approximately US$43 million.
3 Note that energy sources for electric generators in Indonesia are still mostly oil products.
4 In the case of Indonesia, $CO_2$ is the most predominant greenhouse gas.
5 It is a scheme whereby financial value is placed on the carbon stored in forests. Incentives, rewards, and payments are offered to governments, companies, or forest owners in developing countries to reduce emissions from forested lands conditional upon the reduction of deforestation and forest degradation. REDD+ goes beyond deforestation and forest degradation to include the role of forestry in conservation, sustainable forest management, and further development of forest carbon stocks. REDD+ is expected to produce co-benefits of maintaining biodiversity and improving community livelihoods.

## References

Asian Development Bank (ADB) (1997), *Strategy for the Use of Market-Based Instruments in Indonesia's Environmental Management*, Asian Development Bank, Manila.

Ardiputra, I.K. (2004), 'PROPER: Program Penilaian Peringkat Kinerja Perusahaan', Presentation by Kementerian Lingkungan Hidup, Jakarta.

BAPPENAS (2010), *Indonesia Climate Change Sectoral Roadmap: Synthesis Report*, BAPPENAS, Jakarta.

Barber, C.V., E. Matthews, D. Brown, T.H. Brown, L. Curran, and C. Plume (2002), *The State of the Forest: Indonesia*, World Resources Institute, Washington, DC.

Burke, P. and B.P. Resosudarmo (2012), 'Survey of Recent Developments', *Bulletin of Indonesian Economic Studies*, 48(3): 299–324.

Forest Climate Center (2012) 'Indonesia REDD Demonstration Projects: Update March 2012' [on line: http://forestclimatecenter.org].

Hardjono, J. (1994), 'Resource Utilization and the Environment', in: Hal Hill (ed.), *Indonesia's New Order*, Allen and Unwin, Sydney, pp. 179–215.

Hill, H. (2000), *The Indonesian Economy*, Cambridge University Press, Cambridge, UK (Second Edition).

Indonesia REDD+ Task Force (2012), *REDD+ National Strategy*, Indonesia REDD+ Task Force, Jakarta.

International Energy Agency (IEA) (2003), *World Energy Investment Outlook 2003*, IEA, Paris.

International Energy Agency (IEA) (2006), *World Energy Outlook 2006*, IEA, Paris.

International Energy Agency (IEA) (2007), *Statistics & Balances*, IEA [on line: www.iea. org/stats/index.asp].

Jotzo, F., I.A.P. Resosudarmo, and D.A. Nurdianto (2009), 'Climate Change and Development in Eastern Indonesia', in: B.P. Resosudarmo and F. Jotzo (eds), *Working with Nature against Poverty: Development, Resources and the Environment in Eastern Indonesia*, ISEAS, Singapore, pp. 248–267.

Koesnadi, H. (1989), *Environmental Legislation in Indonesia*, Gajah Mada University Press, Yogyakarta.

Ministry of Finance (2009), 'Ministry of Finance Green Paper: Economic and Fiscal Policy Strategies for Climate Change Mitigation in Indonesia', Ministry of Finance and Australia Indonesia Partnership, Jakarta.

Pelangi Energi Abadi Citra Enviro (PEACE) (2007), 'Working Paper on Indonesia and Climate Change: Current Status and Policies', PEACE, Jakarta.

Persatuan Perusahaan Air Minum Seluruh Indonesia (PERPAMSI) (2004), 'Informasi Perbandingan Daftar Tarif Air dan Biaya Operasi Sistem Branching 2004', PER-PAMSI, Jakarta.

Resosudarmo, B.P. (2003), 'River Water Pollution in Indonesia: An Input–Output Analysis', *International Journal of Environment and Sustainable Development*, 2(1): 62–77.

Resosudarmo, B.P. and A. Kuncoro (2006), 'The Political Economy of Indonesian Economic Reform: 1983–2000', *Oxford Development Studies*, 34(3): 341–355.

Resosudarmo, B.P. and Y. Vidyattama (2007), 'East Asian Experience: Indonesia', in: A.M. Balisacan and H. Hill (eds), *The Dynamics of Regional Development: The Philippines in East Asia*, Edward Elgar, Cheltenham, UK, pp. 123–153.

Resosudarmo, B.P. and A.A. Yusuf (2009), 'Survey of Recent Developments', *Bulletin of Indonesian Economic Studies*, 45(3): 287–315.

Resosudarmo, B.P., A. Alisjahbana, and D.A. Nurdianto (2011a), 'Energy Security in Indonesia', in: L. Anceschi and J. Symons (eds), *Energy Security in the Era of Climate Change*, Palgrave Macmillan, Basingstoke, UK, pp. 161–179.

Resosudarmo, B.P., F. Jotzo, A.A. Yusuf, and D.A. Nurdianto (2011b), 'Challenges in Mitigating Indonesia's $CO_2$ Emission: The Importance of Managing Fossil Fuel Combustion', CCEP Working Paper No. 1108, Centre for Climate Economics and Policy, Crawford School of Economics and Government, Australian National University, Canberra, Australia.

Resosudarmo, I.A.N.P., B.P. Resosudarmo, and B. Isham (1997), 'The Indonesian Clean River Program (PROKASIH) as Perceived by the People Residing Along the Rivers in Jakarta', *Indonesian Journal of Geography*, 29(74): 47–64.

Rock, M.T. (2000), 'Using "Green Taxes" to Increase Revenues and Improve Environmental Management in Local Government Following Decentralization', EPIQ-USAID Internal Report.

Sunderlin, W.D. and I.A.P. Resosudarmo (1996), 'Rates and Causes of Deforestation in Indonesia: Towards a Resolution of the Ambiguities', CIFOR Occasional Paper, No. 9(E), Center for International Forestry Research, Bogor.

Wheeler, D. and S. Afsah (1996), 'Going Public on Polluters in Indonesia: BAPEDAL's PROPER PROKASIH Program', East Asian Executive Report, World Bank, Washington, DC.

World Bank (2003), *Indonesia Environment Monitor 2003*, Special Focus: Reducing Pollution, World Bank, Washington, DC, and Jakarta.

World Resource Institute (2007), 'Climate Analysis Indicator Tools. CAIT Version 4.0' [on line: www.wri.org/tools/cait/].

# Conclusion

## Green fiscal reform and future challenges for East Asia

*Soocheol Lee and Kazuhiro Ueta*

This book aims to examine the situation and issues related to environmental tax reforms (ETR defined by the European Environmental Agency, EEA) currently underway in EU member states and East Asian countries and to broader green fiscal reforms (GFR), and to explore the desirable directions for institutional reforms aimed at realizing a sustainable low carbon economy in these regions. EU member states as well as the European Commission have promoted institutional reform in this area for nearly two decades, and East Asia has recently accelerated relevant institutional reform. Together, both of these regions can learn the lessons necessary for institutional reform by sharing their experiences.

East Asia is composed of a number of countries at different levels of industrialization. Countries like Japan, South Korea, and Taiwan became industrialized rapidly in as little as several decades (as opposed to the several generations it took for EU countries), countries like China are in the midst of rapid industrialization, and countries like Indonesia and Vietnam have only a short history of industrialization. Compared to countries in the EU, greater differences exist between the countries of East Asia in regard to the level of economic and social development/maturity and the degree of progress in "greening" fiscal measures.

Against this backdrop, advanced industrialized countries in East Asia such as Japan and South Korea have been rapidly implementing institutional reforms related to their interest in greening fiscal measures. The decision by Japan to implement a carbon tax in October 2012 and the decision by South Korea to start an emissions trading system in 2015 represent the progress made. However, with regard to the carbon tax in Japan, the tax rate is very low (about 300 JPY per $CO_2 \cdot t$) and tax revenue is used to provide subsidies that promote carbon emission reduction. The carbon tax in Japan is thus not intended as ETR defined by EEA (Lee, Pollit & Ueta, 2012). In the case of South Korea, there has been no official action yet to reform the energy tax system and to make the tax rates more consistent with energy products' $CO_2$ intensities. In Taiwan, the government was preparing to introduce a carbon tax but is facing difficulties in realizing it due to strong political opposition. China and other developing countries in East Asia have started discussions in recent years on the introduction of a carbon tax. It is notable that China, the world's biggest source of greenhouse gases, announced recently to introduce a carbon tax in the near future, although uncertain about the timetable and the rate.

In the 2010s, there has been some progress on carbon pricing by ETS in East Asia. Tokyo started an emission trading scheme in 2010, the first metropolitan ETS in Asia. South Korea and Vietnam have approved plans for implementing a national emissions trading scheme. China established trans-regional trading schemes, in Beijing, Shanghai, etc. in 2011, in transition to a national scheme by 2015. Evolution of ETS in East Asia will play a key role in carbon emission reduction and development of low carbon technologies in this area, creating one of the largest carbon market in the world.

ETR includes not only reform of carbon and energy taxes but also reform of taxation on environmental pollutants and resource use that are not related to energy. In East Asia, to deal with industrial pollution, urban environmental issues, waste problems, and various other environmental problems facing the region, some countries have implemented policy measures that take advantage of market mechanisms, such as imposition of taxes and fees on pollutants, resource use, and waste discharge. Environmental measures like subsidies to environmental protection facilities of local governments and private companies supported by revenues from such taxes have clearly played a role in those East Asian countries with a weak fiscal base. In addition, carbon taxes as well as other environmental taxes can spur innovation and also contribute to the diffusion of new technologies and practices (OECD, 2010).

In South Korea, the proportion of these taxes and fees to GDP or overall taxes is particularly high relative to the average for EU countries because the government charges a variety of environmentally related taxes and fees – Korea 2.4 per cent, Japan 1.6 per cent, as compared to OECD average of 2.2 per cent (data for 2011 from OECD). However, the tax and fee rate is not so high. It is difficult to evaluate that they have provided proper incentives for emission reduction. Also, revenue from these environmental taxes and fees is being regularly used for financing environmental infrastructure and is therefore not in accordance with the definition of an ETR as put forward by EEA but which corresponds to the underlying concept of an environmental fiscal reform or green fiscal reform as for example promoted by the World Bank (2005). However, through a reexamination of the tax rate and the use of revenues, these kinds of taxes and fees can be an important factor in ETR. These taxes and fees whose revenues have been used as subsidies to environmental protection facilities provide implications for developing countries in East Asia that urgently need to deal with air and water pollution and other types of environmental pollution without sufficient financial support and thereby addressing and solving the most urgent domestic environmental problems.

Along with environmental tax reform, green reform of fiscal expenditure is also important. Countries should provide companies with stronger incentives to pursue the innovation of low carbon processes or low carbon products, while shifting the focus of expenditure from subsidies for fossil and nuclear energy towards support for renewable energy. Japan and South Korea have increased government expenditure in these fields as part of their green technology strategy and low carbon green growth strategy, respectively. Other Asian countries like

China and Taiwan are also showing increasing interest in more efficient energy use and in developing new industries. This interest can be a factor in improving the political acceptability of green fiscal reform (Mori, 2012).

Green fiscal reforms in East Asian countries will be significantly affected by their energy policies. Because of the severe accident at the Fukushima Daiichi Nuclear Power Plant and the subsequent decline in public support for nuclear power generation, the scenario in which Japan reduces or abandons nuclear power generation by 2030 has become inevitable.

In the near future, it is vital for East Asian countries to create an institutional system that properly assesses the cost of nuclear power generation through the process of strengthening nuclear safety standards, reevaluating the risks associated with nuclear power generation, and reexamining the relevant damage compensation laws. Also, many East Asian countries like Indonesia provide assistance for fossil energy use in various forms, for instance, a direct subsidy for gasoline or an indirect subsidy such as one for parking in urban areas. These countries need to promote a shift away from their fossil fuel dependence by removing such subsidies which also freeze up scarce financial resources.

Concrete steps have been implemented in recent years in many European countries in overcoming the fiscal deficits by mainly relying on spending cuts implemented in austerity packages. But these spending cuts were accompanied by raising additional tax revenues also from environmental taxes (Speck, 2013). Speck, however, stresses in our book that the rationale for environmental taxes should be to create environmental benefits, which are in the case of carbon taxes and the reduction of GHG emissions, and to provide incentive of shifting to less carbon-intensive energy products.

The EU's experiences in environmental tax reform show that the negative economic impact of the resulting higher prices has been minimal. Ekins and Speck (2010) perform an *ex-ante* analysis to investigate environmental tax reform and show that, in the medium to long run, the reform not only increases firms' investment in carbon reduction, but also positively affects employment and GDP. In other words, the analysis revealed the existence of a double dividend. Model analysis for studying environmental tax reform in East Asian countries in this book suggests that the double dividend is realized, but is different in size. Ekins pointed out, in our book, a future challenge is how a country mitigates the negative impact of an energy price increase on income redistribution to the low-income class. The response to this issue will be politically important for promoting environmental tax reform in East Asian countries.

In East Asian countries, the greatest obstacle to green fiscal reform, including environmental tax reform, is low awareness about the harm that the use of fossil energy causes to the environment, also to the health of the population and about the risks associated with nuclear energy. There is also a lack of understanding of the concept of internalizing negative externalities through economic measures, or, in other words, a lack of understanding of environmental economics itself. The reason why some European countries have achieved a certain level of success in environmental tax reform is that the public has been willing to accept the pain of small energy cost increases for the sake of the environment and the welfare of future generations. Moreover, political leadership has made possible policies involving such energy cost increases. GFR is a challenge that is difficult to achieve based on a short-term perspective. Necessary first steps are to present to the public a long-term vision for greening fiscal measures in an easy-to-understand manner and to gain public support.

In promoting green fiscal reform in East Asia, it is necessary to distinguish between the lessons to be learned from the experiences of the EU and its mistakes to be avoided. This book makes a significant contribution in that researchers from the EU, which over many years have built an institutional foundation for green fiscal reform, and East Asia, which is rapidly promoting institutional reform, present lessons necessary for realizing institutional reform in the respective regions through discussions based on mutual understanding.

## References

Ekins, P. and S. Speck eds, (2011), *Environmental Tax Reform (ETR): A Policy for Green Growth*, Oxford: Oxford University Press.

Lee, S.C. ed. (2010), *Environmental Charges in East Asia: Factors and Barriers for Institutional Evolution*, Kyoto: Showado (in Japanese).

Lee, S.C., H. Pollitt, and K. Ueta (2012), An Assessment of Japanese Carbon Tax Reform using the E3MG Econometric Model, *Scientific World Journal*, 2012, ID 835917.

Mori, A. ed. (2012), *Democratization, Decentralization and Environmental Governance in Asia*, Kyoto: Kyoto University Press.

OECD (2010), *Taxation, Innovation and the Environment*, Paris: OECD.

Speck, S. (2013), Carbon Taxation: Two Decades of Experience and Future Prospects, *Carbon Management*, 4(2), 171–183.

World Bank (2005), *Environmental Fiscal Reform: What Should Be Done and How to Achieve It*, Washington, DC: World Bank.

# Index

Page numbers in *italics* denote tables, those in **bold** denote figures.